Green
Alaska

DREAMS FROM THE FAR COAST

Green
Alaska

Nancy Lord

COUNTERPOINT
WASHINGTON, D.C.

First Counterpoint paperback edition 2000

Illustrations courtesy of the Smithsonian Institution Archives and the Library of Congress.

Portions of this book appeared in *The Seattle Times* and *Alaska Magazine*.

Library of Congress Cataloging-in-Publication Data
Lord, Nancy.
    Green Alaska : dreams of the far coast / Nancy Lord.
       p. cm.
    ISBN 1-58243-078-0 (alk. paper)
    1. Alaska—Description and travel.   2. Lord, Nancy—Journeys—
Alaska.   3. Harriman Alaska Expedition (1899)   4. Natural history—
Alaska.   5. Alaska—Environmental conditions.   I. Title.
F910.5.L67   1999
917.9804'51—dc21
                                                                                99-11194
                                                                                    CIP

Printed in the United States of America on acid-free paper that meets the American National Standards Institute Z39–48 Standard.

COUNTERPOINT
P.O. Box 65793
Washington, D.C. 20035-5793

Counterpoint is a member of the Perseus Books Group.

10 9 8 7 6 5 4 3 2 1

Contents

III.

Washington, D.C.

LIST OF ILLUSTRATIONS

ACKNOWLEDGMENTS

Green Alaska: Dreams from the Far Coast is an imaginative work based largely on the Harriman Alaska Expedition of 1899 and writer John Burroughs's role as expedition historian. It is not a work of scholarship, although it relies on the work of many scholars.

Primary texts consulted were Burroughs's "Narrative of the Expedition," which begins volume one of the thirteen-volume *Harriman Alaska Series* (and is also included, as "In Green Alaska," in *The Writings of John Burroughs*, vol. 13), and the rest of volumes one and two of the expedition reports, most easily found in *Alaska: The Harriman Expedition, 1899* (reprint edition by Dover Publications, 1986). Additional useful texts were George Bird Grinnell's *Alaska 1899: Essays from the Harriman Expedition* (University of Washington Press, 1995), with introductions by Polly Burroughs and Victoria Wyatt, and John Muir's "Cruising with the Harriman Alaska Expedition," included in *John of the Mountains: The Unpublished Journals of John Muir* (University of Wisconsin Press, 1979).

Of great help was the one scholarly book written entirely about the Harriman Expedition, *Looking Far North: The Harriman Expedition to Alaska, 1899* (Viking Press, 1982), by William H. Goetzmann and Kay Sloan.

I depended for my knowledge of Burroughs on his twenty-seven volumes of published works, as well as several biographies of him, including Clara Barrus's *The Life and Letters of John Bur-*

roughs (Houghton Mifflin, 1925), Elizabeth Burroughs Kelly's *John Burroughs: Naturalist* (Exposition Press, 1959), Perry Westbrook's *John Burroughs* (Twayne Publishers, 1974), and Edward J. Renehan Jr.'s *John Burroughs: An American Naturalist* (Chelsea Green, 1992). *Birch Browsings: A John Burroughs Reader* (Penguin, 1992), edited and with an introduction by Bill McKibben, anthologizes some of the best of Burroughs's nature writing. I recommend as well both Paul Brooks's *Speaking for Nature: How Literary Naturalists from Henry Thoreau to Rachel Carson Have Shaped America* (Houghton Mifflin, 1980) and Frank Stewart's *A Natural History of Nature Writing* (Island Press, 1994) for placing Burroughs's contributions within larger contexts.

I am grateful to the Smithsonian archives for access to the Souvenir Album and related materials.

I thank Patricia Hampl for her early inspiration and advice, the Mershon and Laukitis families and the crew of the *M&M* for making my trip to False Pass both possible and pleasurable, and Nevette Bowen for introducing me to Washington, D.C. Special thanks go to my agent, Elizabeth Wales, for her counsel and encouragement, and to Jack Shoemaker and Trish Hoard at Counterpoint Press for their understanding and vision. As always, I particularly value the companionship, capabilities, and good humor of my partner, Ken Castner.

I very much appreciate the support of the Island Institute in Sitka, Alaska, the Anderson Center for Interdisciplinary Studies in Red Wing, Minnesota, the Virginia Center for Creative Arts, Fundación Valparaiso in Mojacar, Spain, and the Alaska State Council on the Arts—all of which provided me with secluded time and enthusiastic fellowship for work on this project.

A number of generous friends have reviewed some or all of the text and offered helpful comment. These include Buck and Shelly Laukitis, Tom Kizzia, Karen Wessel, Wendy Erd, Kim Cornwall, and David Roberts.

Our completed party now numbered over forty persons besides the crew and the officers of the ship (126 persons in all), and embraced college professors from both the Atlantic and Pacific coasts—botanists, zoologists, geologists, and other specialists, besides artists, photographers, two physicians, one trained nurse, one doctor of divinity, and at least one dreamer.

JOHN BURROUGHS,
"Narrative of the Expedition"

One of the oldest dreams of mankind is to find a dignity that might include all living things. And one of the greatest of human longings must be to bring such dignity to one's own dreams, for each to find his or her own life exemplary in some way.

BARRY LOPEZ,
Arctic Dreams

The Dreamer's name Qatsitsexen, literally means "the one with prescient dreams," and the Dreamer actively sought dreams through which to provide guidance to the people. In traditional Kenai Peninsula Dena'ina thought, dreams were perceived quite differently from Western culture. In psychoanalytical thought dreams are generally considered to be an individual's subconscious expression of an event of one's past, and, because they are considered to be beyond conscious control and vague, take on an illusional or surrealistic dimension. Dena'ina dreams were not about the past or present, but about the future, and represented, not an illusion, but the reality of something that surely would happen.

ALAN BORAAS AND DONITA PETER,
"The True Believer Among the
Kenai Peninsula Dena'ina"

Environmentalists have been looked on as the dreamers of the world, where in fact they're the realists.

E. O. WILSON,
Audubon

In dreams begins responsibility.

WILLIAM BUTLER YEATS,
Responsibilities

Imagine This

Here is a man, the Bill Gates of a century ago—in 1899 perhaps the richest man in America. Edward H. Harriman, president of the Union Pacific Railroad, can afford to do absolutely anything he wants on the vacation his doctor insists he must take. If it were technically feasible, he might fly to the moon.

The exotic then lay not quite so far away. Harriman leases a large steamship and completely reoutfits it for luxury travel through the waters of Alaska. Not just the southerly inland passage that has become fashionable but all the coast of Alaska, all the way to Bering Strait and the shoreline of farthest Siberia. The water part of the trip, sandwiched between overland journeys by train, shall last the two most light-filled and calm summer months.

In his orderly, take-charge fashion, Harriman supplies the ship with motor launches and canoes, a piano and organ, weaponry for hunting, horses and tents, cases of champagne and the requisite thin-stemmed glasses, a library, the latest audio and visual equipment, steers and sheep, turkeys and chickens, a single milk cow, a derrick to lift aboard the great Alaska brown bear he desires. Also, the necessary personnel consisting of—in addition to the ship's sixty-five officers and crew—doctors and a nurse, a preacher, stenographers and taxidermists, hunters and packers.

The passenger list: Harriman's wife and children and a few choice friends. And then, because he is not only rich but gener-

ous, Harriman invites thirty more guests—the nation's top natural scientists, mainly, but also a few practical engineering types,
some cultural enthusiasts, select writers and artists and photographers. Everyone has his assignment, his field of study. (*His* because they are all—excepting those females among family and
friends—men.) All will participate in the lectures and entertain-
ment aboard ship, in the "floating university."

There may be more to Harriman's dreaming, though historians only speculate. Very likely he looked to Alaska for some kind
of exploitable economic opportunity, some resource to haul
away by the trainload. Possibly he entertained the idea of a railroad line that might circle the world, crossing from Alaska to
Siberia beneath Bering Strait. Such a thing, outrageous as it may
seem, had been discussed.

But this book is not about dreams to circle the world with rail or
to get rich packing off coal and timber. There are other dreams
here, and one is simple, romantic exploration, going where few
had gone before. Only thirty-two years had passed since America had made its questionable Russian purchase, and the great
land was still more mystery than known. Two of the experienced
Alaska hands on the expedition—William Dall and John Muir—
had led such primary roles in previous explorations that animals
and glaciers, respectively, were named for them. All along the
Alaska coast there were distances to be measured, tree rings and
fish streams to be counted, worms to be collected. There were
Indians eating gull eggs and boiled marmot and Eskimos who
paddled skin boats. And there were, all around, strange new vistas to admire, photograph, paint, and describe.

Who would not have *begged* to have been invited on such a
trip, to witness Alaska at the century's cusp, when it and the
world were still so new? When so much was possible? It was the
age of innocence, still ruled by infectious Victorian optimism,

and Alaska in 1899 was—as perhaps it still is—a place of promise.

Picture, among the guests boarding in Seattle, John Burroughs, sixty-two years old, the most venerated nature writer in America, indeed among the most admired of all American writers. In his high, pointy-toed leather boots, and beneath his dark brimmed hat, he looks as jaunty as any of the gentlemen. The white beard that reaches past his lapels enhances the kindly, grandfatherly look that has helped make him a favorite of school children. He waves, in a generalized, politician's way, to the crowds that have gathered to see the ship off. He has already made the long trek cross-country by train, to a strange land where he knows no one but his recently met expedition colleagues. The only fellow traveler with whom he has previous direct acquaintance is "the other John," John Muir, who—rowdy with stories about glaciers and a favorite dog—has just joined the party. Already Burroughs has confided to his private journal, "Have I made a mistake in joining this crowd for so long a trip? Can I see nature under such conditions?"

Because nature is what he means to see. He has made a fine art out of observing the tiniest and tenderest moments of bird feather preening and honeybee nectaring, and his eleven books to date fill reading rooms and bedside tables all across the country. He more than any person has fed the newly popular American passion for nature study and idolatry, the national "back-to-nature" movement that is in full swing. The song of the hermit thrush, the "liquid drapery" of waves on a shore, the beauty of ordinary corn tassels, these all have been brought to readers through Burroughs's finely articulated, exact, and appreciative language. There is no Thoreauvian discomfort in Burroughs's writings; no metaphor, allusion, or lesson—only delight.

Burroughs's vision is Romantic, mythologizing a harmonic rural past. Nature is soothing, good for the soul and the appetite—like strolls down country roads. Burroughs's sight takes in the old worn hills of the Catskills, fields and vineyards, the slow current of the Hudson. His natural world is a civilized and homey place. One of his fellow naturalists, Dallas Lore Sharp, will say with pinpoint accuracy, "Showing somewhere along every open field in Burroughs' book is a piece of fence, and among his trees there is always a patch of gray sloping roof."

Never before has Burroughs traveled west of the Mississippi. Never before has he experienced purely wild country.

His job: expedition historian, the one to write the official trip narrative.

And here are the others, awaiting departure. Find among them:

C. Hart Merriam, head of the U.S. Biological Survey and a noted mammalogist. As leader of the scientific party, he has selected most of the participants and bears some responsibility for keeping them happy and in line. Despite his rank, he is very much the field biologist—quick to get ashore to set his mouse traps, willing even to extend his scientific curiosity to charging into a herd of sea lions. He is also a decent amateur photographer, with an artful eye and a sense of historical importance.

George Bird Grinnell, editor of the outdoor magazine *Forest and Stream* and founder of the Audubon Society. Asked along for his twenty years' study of Montana's Blackfoot Indians, he is expected to contribute his observations of Alaska's Native cultures. He brings a clear-eyed, practiced pessimism to his examinations of social and economic upheaval.

The fortunate Edward Curtis. A year before, as a young fashion photographer with a passion for mountains, he helped out a lost hiking party on Mount Rainier. Two members of that party—Merriam and Grinnell—remembered him when it came time to find an expedition photographer. Although the main body of his work on the Harriman trip will be more than 5,000 landscape photos, he will also learn something from Merriam about photographing Alaska's Natives, and under Grinnell's influence will go on to devote his professional life to documenting what he'll consider the vanishing race of American Indians.

William Dall, one of the old men of the expedition. Renowned as the first American naturalist to study in Alaska, he remains an authority on a great number of things, from history to pale-ontology and now to his current specialty, mollusks.

John Muir, great talker and intrepid conservationist, enthusiast of glaciers. He is irreverent about the expedition's hunting and collecting, and he will say without embarrassment, "Why, I am richer than Harriman. I have all the money I want and he hasn't."

Many more, of course, some of them famous still: bird artist Louis Agassiz Fuertes, glaciologist Grove Karl Gilbert, geographer Henry Gannett. The eight-year-old in the sailor suit: Averell Harriman—future ambassador, governor, husband to a society wife.

A hundred years later, in cruiseship-swarmed, oil-rich Alaska, the Harriman expedition lies embedded in our maps, our history, the base of what we know. Everywhere I turn, I run into the references: Harriman Fjord, Mount Curtis, Cape Fox Village, my town viewed as a sand spit, a much-reproduced photo of John Burroughs and John Muir together, glacier theory, bear stories, a footnote about Eskimos and whaling ships.

When I have a chance to travel by boat—a salmon tender—from Cook Inlet to the end of the Alaska Peninsula, I think, *the Harriman Expedition went that way*. It was only one segment of their long journey, but nevertheless, they went that way: steaming down Shelikof Strait, all the length of the peninsula to the Aleutians and beyond.

Into my duffle, I slip between wool hat and raingear my great-grandmother's 1904 leather-bound copy of *The Writings of John Burroughs*, volume 13, "In Green Alaska," his expedition account.

Burroughs, this man who called himself a dreamer, who has quietly faded from our libraries and anthologies, feels kin to me. I grew up, before moving to Alaska, on his settled east coast; I know his landscape. His books, their pages safeguarding three generations of family photos, rested on my grandmother's and then my parents' shelves before coming to live with me. Burroughs is familiar in the family sense. But I like him, too, because he knows how to look carefully at things and to choose his words with equal care; his prose has always seemed to me cleanly written, delightfully observant, quietly humorous. And I like him because he knows about loving places.

The country we will pass is thinly inhabited and little developed—as unchanged today as any of the coastline the Harriman expedition followed. We will be passing it at a speed similar to the Harriman ship, our sea-level views of mountain peaks and watery depths essentially the same as those witnessed by Burroughs from his deck chair and porthole.

What better place and time could there possibly be to contemplate this amazing land, an Alaska that still lives in most people's minds, if they think of it at all, in an imaginary realm? Let me look; let me see. Let me read. In the great silence that disregards the steady roar of our diesel engines and the pounding of waves beneath the hull, I want to dream about what is, was, and may be, not just along one stretch of coast, but throughout Alaska, the north, the world.

PART I

Slant

As we start out of Kachemak Bay, the halibut charter boats are heading in, skimming over the flat surface, leaving wakes like contrails. Dozens and dozens of them returning from the outer bay and Cook Inlet, all zeroed in on the harbor at the end of the Homer Spit. They advance like a naval assault, sure and speedy and exactly to target.

The sports fishermen, sun-baked, feet still rolling with sea swells, will pose in the harbor beside their strung-up fish, "chickens" in the twenty-pound range, and bigger, bigger than themselves—200-, 300-, 350-pound fish. Last year, the winning fish in the local halibut derby weighed 379 pounds.

My partner, Ken, who is taking this hundred-foot crab-boat-modified-to-be-a-salmon-tender out to the Area M salmon fishery for the summer, puts the *M&M* on autopilot so he can help the crew finish tying down the on-deck load.

I raise my brows. "With all this traffic?"

Ken says, "They usually get out of our way."

The Spit, receding behind us, looks ponderous, a mass of commerce piled onto nothing more than a sandbar. There is the square-sided cannery building where I got my first Alaska job cracking crab legs so many years ago. There is the hotel, aptly named Land's End, where Ken, also early in our history, worked the night desk. White fuel tanks gleam, cranes lift over the fish dock, and then there are all those log piles and a mountain of yellow wood chips waiting to be floated off to Japan and Korea.

Barnroof tourist shops—loaded with smoked salmon samplers
and moose nugget jewelry, T-shirts and hats, stuffed puffins and
sand dollar refrigerator magnets—jam the elevated boardwalks.
Most every inch that doesn't lie beneath a building or a log pile
has become the property of elephantine motor homes, enough
to encircle a mid-size city.

Inland from the Spit, the Homer bench and hills glitter: sun
reflecting from windows and metal roofs. As I watch, a plane
climbs steeply from the airport and swings north, toward Anchorage.

Such a lot of so much, I think, with both fondness for my
hometown and a panging regret, paradise lost. I have seen the
town go from one halibut charter business to this enormous
fleet, from dirt streets to a speedy bypass and a McDonald's. The
realization comes to me as a shock: I have lived here for fully
one-quarter of the time that has passed since the Harriman expedition stopped by. No wonder so much has changed.

"There was nothing Homeric in the look of the place," Burroughs wrote in 1899, by which I suppose he meant there was
nothing very classy about the town, nothing that its human inhabitants had done to elevate it above its own raw nature. Harriman's family and friends were perhaps Homer's first tourists,
though few of the party actually stepped ashore; they were visited instead by a Dr. Gunning and a Captain Ray "who is exploring the coal," and they left off mail. In this, their disinclination
to debark, they are not unlike today's cruiseship tourists, most of
whom remain where the food is already paid for; those who do
venture over the dock tend to complain about the piles of logs
and the fishy smells.

The logs offend me, too, for a reason that has nothing to do
with the immediate aesthetics. The fate of Alaska's forests is one
the Harriman Alaska Expedition (H.A.E.) miscalculated altogether. The forestry expert aboard judged Alaska's trees inferior

and difficult to harvest, as well as too far from markets, and promised they would be left untouched except for local use. These many years later, the lush rainforests of southeast Alaska have been taken down, and now the industry is rapidly shaving our slowest-growing boreal woods of their ancient, spindly spruce.

I slant my eyes and try to see the sand spit and green hills, plain, as Burroughs did. It was the end of June, a hazy day, when the *George W. Elder* anchored up inside the bay and Burroughs noted that the entire "hamlet" consisted of four or five low wooden buildings on the end of the spit. The town, such as it was, had only come to exist with a name and post office three years before, when a mining company that exploited investors instead of gold briefly made it its headquarters. The fifty men and one woman named the settlement after their leader, Homer Pennock, but never spent a winter, and by 1899 they had relocated to the fevered Klondike.

The one woman with all those men wrote many years later, "I think it was the most desolate spot I have ever seen."

We motor toward the open inlet while I continue to look slant behind, the spit getting lower, the town shrinking into the larger landscape. *Desolate* is not a word I would ever have put to this place, however unwelcoming those early wayfarers and then Burroughs and his fellow expeditioners may have found it. Edit out the fuel tanks, the roads clogged with motorhomes, the boardwalks and shops and all the industrial development, and the place becomes, for me, a long beach littered with driftwood, waving grasses hiding the nests of eiders, luxurious green hills beyond. Even in winter, or especially in winter, I find the soft light and shadows and all the edges between land and sea achingly lovely. Sky. Water. Beatific space. Not desolate but desirable—I would say *Edenic*. I want to believe that a hundred years ago I would have fallen all over myself in love with it.

George W. Elder:

A NATURAL HISTORY

G *eneral description*
From afar the *George W. Elder* can be recognized by its plume of black coal smoke and by its whistle, heard especially in fog. Twin masts, fore and aft, extend to twice the height of the squat central smokestack. The square-bowed iron body is dark below, white above, with pilothouse and life boats perched on the top, hurricane deck.

Dimensions are a length of 250 feet and a beam of 38.5 feet, lending the ship a sleek proportionality. Beneath the waterline lies more dark hull and a large three-bladed propeller.

Weight is 1,709 tons.

Closer examination will identify rigging, portholes, and a cargo of free-moving human passengers who pass back and forth across the decks. Two young boys are dressed in nautical suits.

Internally, stairs and corridors lead to staterooms, galley and dining room, library, salon, crew's quarters, engine room, stowage space, coal bunkers and furnaces, livestock stalls and pens. Casks are filled with fresh water. The library holds five hundred volumes, including most of what has been written about Alaska.

Origins
This specimen originated in 1877 with the Oregon Steamship Company, and was known at the time as one of the fastest and

most modern cruise liners on the West Coast (five-day round-trip from Portland to San Francisco).

In 1888 it served as the mail steamer serving southeast Alaska. Education agent Sheldon Jackson had it deliver two totem poles from Metlakatla to Sitka for a museum he was starting there.

As a charter boat for the Harriman expedition, it was entirely overhauled and refurbished at Harriman's expense.

Common name
George W. Roller, for the way it rolls in moderate to heavy seas. Passengers are frequently seasick.

Behavior
Like all of its propeller class, this specimen is powered by a steam engine that turns a propeller (in contrast to a paddle wheel). The fuel is coal, and large black plumes of smoke are exhausted. The average speed is 12 knots. When at anchor, the ship is so quiet a passenger may hear the songs of birds from ashore.

The ship's captain relies for navigation on charts, compasses, soundings by leadlines, the echoing of the ship's whistle, figuring distances by traveling for known times at known speeds, and his eyes, ears, and nose. Nevertheless, the ship frequently, like others in its class, runs aground.

Song
A long whistle, used chiefly to call passengers back aboard after a stop.

Also, a variety of sounds from its waxed-record graphophone, used to entertain passengers and sometimes played from the deck at full volume to greet or send off welcoming or departure parties in port. Two recordings made in Sitka and replayed along the way are traditional Tlingit songs and Sitka's all-Tlingit brass band blasting out "Yankee Doodle" and "Three Cheers for the Red, White, and Blue."

Additional vocalizations emanating from the salon include hymn-singing, college pep cheers, and repeated chants of *What's the matter with _____? He's all right! Who's all right?*

Future prospects
In 1905 this specimen will strike a rock in the Columbia River and be submerged for more than a year before being raised and returned to service. In 1907 it will rescue passengers from a collision of two other steamers, in which 100 lives will be lost. Last known sighting: a 1918 newspaper account will have "the ancient steamer" bound for Puget Sound from South America with a cargo of fertilizer nitrates.

The Price of Otters

We are still in Kachemak Bay when we spot our first sea otter, riding high in the water on the float of its air-filled fur. It rolls from its back to its side and paddles lazily just out of our path, then turns belly-up again to bob in the wake of our bow thrust. We are close enough for mutual regard, direct full-face eye contact, and I catch in its expression something I interpret as part curiosity, part disdain. It sees more of my kind than I see of its, and so the greater interest, the novelty, is mine. I examine the gray that blooms in its old-person face, the hard shine of its beaded eyes, the one paw held casually over its chest in a gesture that looks oddly effete.

When I first came to live at the edge of the bay, there were no otters here. Later, in small, shy numbers, they migrated in along the bay's south side. They were a treat to see, preening air into their fur, or prying crabmeat from shells they broke and sucked and tossed aside like picked bones. Gradually they spread around and across the bay, rafting into tribes, crying out in storms, the females sometimes bleeding from chewed-on noses they acquired in a cruelly vigorous mating process. They thinned out whole coves of juvenile crabs, mussels, clams they broke open with rocks they carried with them. They ate the sea stars and urchins that eat the kelp, and so they helped bring back the kelp forests that provide food and shelter for so much else. I learned to call them a "keystone" species, one that holds up the roof over the rest of us; where there are otters, there will be other species, diversity, some measure of environmental health.

The otter we've passed is now just a speck on the water, still riding high on its thick and buoyant ever-so-treasured coat of fur. I have never touched an otter, but I *have* stroked a pelt displayed in the museum in Homer for that purpose. Otter fur is soft beyond imagining—softer than any kitten, any mink coat or rabbit hat, any crushed velvet. It is, in fact, the finest, densest fur in the world, up to half a million individual hairs in every square inch. The thick fur, and the air trapped in the fur, means that an otter's skin never gets wet. When I touch the museum's pelt, the feel of it on my fingertips, against my palm, is like heat.

Otters were the beginning. They opened Alaska to the first wave of exploitation from without, drawing eastward the Russian fur traders who enslaved the Aleut people and forced them to hunt otters to the brink of extinction.

It is hard for me to imagine, given the length of Alaska's crannied and treacherous coast—that long arc of otter habitat from southeast to the Aleutians and beyond—that the machinery of slaughter could have been so efficient. There were finally so few sea otters left that hunting them became unprofitable. In 1910, the last year before otters were protected by international treaty, a steamer with a twelve-boat crew cruised the Aleutians all summer for a total take of only four otters.

In 1899, Burroughs was unlikely to have seen an otter; they would have numbered then no more than a couple of thousand in all the world, the canny survivors wary of anything that looked like a ship or hunter. Burroughs wrote nothing about otters, but he did report, on the visit of the H.A.E. to the fur seal breeding grounds in the Pribolof Islands, that "those of our party who had been there before, not many years back, were astonished at the diminished numbers of the animals,—hardly one tenth of the earlier myriads." He did know something about the pillaging of Alaska's sea mammals for their fur.

Burroughs did not write a word about the otter pelt purchased for $500 by Harriman at a stop in Yakutat Bay. Perhaps he didn't know? I think he did know, and was embarrassed, by the display of wealth and by what Harriman apparently didn't recognize or care about himself—his own culpability in the species' near extinction.

Supply and demand. Five hundred dollars is not the most that had been paid, or may yet be, for an otter. The early Chinese mandarins, shopping with the Russians, were said to spend the equivalent of $5,000 for a single skin. In 1910, skins fetched $2,000 each on the London market. I do some numbers: Five hundred dollars at a time when $2.50 was a working man's daily wage would equate to something like $20,000 today.

In 1989, when the *Exxon Valdez* dumped its oil in Prince William Sound, an estimated 3,905 sea otters died in it—from its ingestion as they madly preened, and from hypothermia when their soiled coats could no longer keep them warm and dry. Exxon paid for the rescue and cleaning of 344 otters, of which nearly two-thirds "recovered" well enough to be returned to the wild. The cost per animal "saved"—for rescue, transport, vet care and medicine, a seafood diet, swimming pool facilities, and constant monitoring—was $80,000.

Today perhaps 100,000 sea otters live in Alaska's waters. Commercial hunting of them has been forbidden since 1911, but under federal law Alaska Natives may take otters to use the furs themselves or make them into handicrafts for sale. This recent reinstatement of traditional usage was not without controversy. Most people understand that Native hunting of marine mammals—walruses, seals, polar bears, bowhead and beluga whales—is grounded in long cultural traditions, even when those traditions were interrupted or carried on sublegally. With sea otters, though, the thread stretched farther and thinner all the way back to a time before the first Russian fur traders arrived and changed everything forever. There were no ethnologists studying the Aleut those many generations ago, and the wisdom of the people about living with otters was lost with the otters in the enslavement process.

Last year, 605 otters were reported taken by Native hunters in Alaska, a number said to pose no danger to their continued recovery and expansion. Perhaps 50 were shot right here on Kachemak Bay. There are no limits on how many can be taken, only a general requirement that the take not be wasteful.

Any tourist can walk into Homer's museum gift store and buy a handcrafted pair of beaded sealskin mukluks trimmed at the top with ankle-warming otter fur, for $500. The same tourist can take a boat ride and find the real live thing, that otter lost now in our blue distance, bobbing like a cork.

CHEER

It's caught in my head, repeating over and over again like a ridiculous line from an old summercamp song.

Who are we? Who are we?
We are, we are, H. A. E.!

And who are *we*? Not an expedition, we, but five people on the way to work, and me with the notebook. Ken does this tendering some years during the weeks when the salmon fishing he and I do together slows; he is good with boats and at going without sleep, and he entices the fleet to deliver to him with fillets of smoked salmon he prepares on the ship's back deck. His deck crew, Dan and Riley, are a pair of young Ishmaels, off to see the sea and earn a few dollars, to the rhythm, always, of loud grunge music. Mei, a tiny woman from Hawaii, cooks, cleans, and soon will be writing "fish tickets," the all-important records of salmon weights and species received from each boat. Vince is not part of the crew but a "fish tech" from the cannery who, like me, is along for the ride; on the fishing grounds he will switch among tenders to check on fish quality and see that species are properly sorted and paid for.

Me, with the notebook—I am primarily a passenger, without specific responsibility. When we reach the grounds I will hop off to see friends who live and fish in that far-western land, and then I'll fly home. My only task at the moment—as we pass a landing craft with, of all things, a car aboard, and two people jigging off

the side—is to try to imagine Burroughs singing out that silly
H.A.E. cheer with the rest. I think he's a bit too serious for such
a thing. He would much rather be curled up in his bunk, pen-
ning an ode to a thrush or a letter to his son at Harvard. But then
I remember.

One evening on their cruise the captain sent up a stoker and a
deckhand to entertain Harriman's assembled guests. One sang,
and the other danced on a hatch cover brought into the salon for
that purpose. When they were done, members of the scientific
party, normally given to formal postprandial lectures but not to
be outdone by the ship's crew, took their own turns on the hatch
cover. John Muir did a neat double-shuffle, and then Burroughs
stepped up with a performance of his own.

I picture him, this saintly old man, clasping his hands behind
his waistcoat, and then, to whatever music was offered, lifting
and clacking and clonking his heels, clog dancing away, all his
limbs loose and flowing like water. His head ducks up and down,
and so goes his elegant white beard, floating off his chest with
each jerk of his chin. A small, shy smile catches in the corners of
his mouth. The memory is all in his country boy's bones.

THE TWO JOHNNIES

he first time they met, six years earlier, they already knew one another by reputation. Each admired the other's writings.

Of that first meeting, in New York City, where Burroughs had attended a memorial celebration for Whitman and Muir had been preparing for a trip to Europe, this:

Muir: "He had made a speech, eaten a big dinner, and had a headache. So he seemed tired, and gave no sign of his fine qualities."

Burroughs: "an interesting man with the Western look upon him. Not quite enough penetration in his eyes."

The second time, three years after their first meeting and three years before they'd each receive an invitation to the Harriman expedition, Muir visited Burroughs at Slabsides, the cabin he'd built on the far side of his Hudson River property.

Burroughs: "A very interesting man; a little prolix at times. You must not be in a hurry, or have any pressing duty, when you start his stream of talk and adventure. . . . He is a poet and almost a seer. Something ancient and far-away in the look of his eyes. He could not sit down in the corner of the landscape, as Thoreau did; he must have a continent for his playground. . . . Probably the truest lover of Nature, as she appears in woods, mountains, glaciers, we have yet had."

Muir's expedition poem, concerning Burroughs:

> . . . *Flat and limp on a deck chair*
> *With naught but his nose bare . . .*

The expedition women, Muir complained, were always hovering around Burroughs, tucking him up with rugs or running to him with flowers and birdsongs. Burroughs was "Uncle John" to the children and some of the younger scientists, "Johnnie" to Muir; together they were "the two Johns" or "the two Johnnies."

Ten years after the expedition, when Burroughs joined Muir to see the Grand Canyon and Yosemite Valley:

Muir to Burroughs: "I puttered around here for ten years, but you expect to see and do everything in four days! You come here, then excuse yourself to God, who has kept these glories waiting for you, by saying, 'I've got to get back to Slabsides.'"

Burroughs to his journal: "See how tender Muir assumes to be toward the animals! Yet he likes to walk over the flesh of his fellow men with spurs in his soles."

On Muir's death, six years before his own:

Burroughs: "A unique character—greater as a talker than as a writer—he loved personal combat and shone in it. . . . I shall greatly miss him."

These two men, destined to be the two great nature writers of their age, were absolute contrasts. East and West, quiet and loud, mild and combative; they praised entirely different Natures. Muir was said to have badgered Burroughs on the expedition, criticizing him for his bovine contentedness back in the East and for failing to speak out for the protection of forests (Muir's latest mission) and other conservation causes. All this he did with rather good humor, apparently—with a fondness that

Burroughs returned. But Burroughs, too, could twist a barb. Describing Muir in his official expedition account, he wrote, "In John Muir we had an authority on glaciers, and a thorough one—so thorough that he would not allow the rest of the party to have an opinion on the subject."

Today the Muir and Burroughs names affix to neighborly sheets of ice in Glacier Bay. On the Harriman expedition, Muir was returning to the bay he'd "discovered" with Indian guides during his Alaska explorations twenty years earlier, and where the prominent Muir Glacier already carried his name. Burroughs Glacier, smaller and less well known, lying just to the west of Muir Glacier, was named for the second John sometime following 1899.

I could do worse for literary traveling companions. My ear is tuned to what they can tell me about their time, about what they knew and saw and believed. Muir speaks easily to us these hundred years later—his famous dictum about finding everything in the universe hitched to everything else very much the message of modern environmentalism. The quieter John is, I think, more difficult to hear but no less knowing.

MIRAGE

The aspects of the sky and atmosphere along the Alaska coast
have a character unlike that of any other part of the United States,
and give an especial interest and charm to the scenery.

WILLIAM H. BREWER,
"The Alaska Atmosphere," *Harriman Alaska Series*

T*he Alaska Peninsula* rises ahead of us, on the far side of Cook Inlet. And rises, and rises. From this distance, the peninsula normally appears as little more than an edge on the horizon. Mirage has lifted the land, doubled and redoubled it, so that it looks this evening like towering high cliffs shimmering across the water.

There's nothing wishful, nothing hallucinatory, about this mirage. It is visual truth, a turning of light through the atmosphere. The air higher over the water is warmer, and so less dense, than air nearer the ocean, and the light passing through the layers doubles one image on top of the other. When summer arrives, land masses normally hidden below the horizon suddenly lift into view, and the thin lines of shores thicken. Hills become mountains.

Burroughs witnessed the same thing. He described islands floating in the air, and capes doubled upon themselves, a snowy mountain range rising over a nearer rocky one so that it "suggested a vast Grecian temple crowning a rocky escarpment."

The magician *Mirage*, he wrote, played tricks with sea and shore all one afternoon.

We think we know something as definite and unchanging as a profile of land, and then we look again and find it twice as tall, as though the earth had opened and spilled out that much more fiery new magma.

I get the language down: superior and inferior. Superior mirages, these raised landscapes. Inferior, those dark watery patches on hot highways; when the warmer air is close to the surface and the cooler air above, the refracting light drops sky to earth. All the lore that goes with thirsty men in the desert. And *fata morgana*—that lovely Italian name, Fairy Morgan of the Arthurian legend, thought to be responsible for the dramatic Mediterranean land-liftings between Sicily and the booted mainland, now referring more generally to any doubled images suspended in the air.

For a long time I watch from the pilothouse window, savoring the light, the liquid look, the airy space that wavers through the height of land. It *is* magic, even knowing the why of it.

Behind us in the bay, out of sight now, there is a small rocky island called Sixty-foot Rock. Before white people named it that, the Dena'ina Athabaskans of this place knew it by a name that translates to "Soles of Feet Waving." I used to study that solid mound of rock and wonder how it came to have so lovely and unlikely a name, and then one day I imagined early Dena'ina looking across the water from low in their boats and seeing the island raised above itself in mirage, as tall as the bottoms of a giant's feet. And waving, waving, in the bent, unsteady light.

1899, a Year

The Treaty of Paris to end the Spanish-American War was ratified by the U.S. Senate, and the United States became a new international power with control over both Cuba and the Philippines. The Boer War began in South Africa, between the Afrikaners who were there sooner and the British who had flooded in after gold was discovered. In India severe drought killed one and a quarter million people. The boll weevil crossed the Rio Grande from Mexico and began its devastating spread through U.S. cotton fields.

Aspirin was perfected in powdered form from coal tar, and cancer was treated with X rays. The Vice President of the United States, Garret Hobart, died in office.

The rich got richer. Rockefeller consolidated his many oil-refining companies into Standard Oil of New Jersey. Carnegie Steel was created by the consolidation of various steel properties controlled by Andrew Carnegie. International Paper formed in a merger of nearly 30 U.S. and Canadian paper companies, Consolidated Edison Company in a merger of New York's Edison Illuminating Company with Consolidated Gas, and American Smelting and Refining Company in a joining of U.S. copper producers. Seven chinook salmon canneries located at the Columbia River's mouth became the Columbia River Packers Association.

The most popular poem in America, "The Man with the Hoe," by California schoolteacher Edwin Markham, protested the plight of the poor farmer. In Chicago, pioneer social worker Jane Addams pushed for the creation of the world's first juvenile courts.

One in 10,000 Americans owned a car. Of these, 40 percent were steam-powered, 38 percent were electric, and 22 percent were gasoline-powered. A Stanley Steamer was driven by F. E. Stanley to the top of New Hampshire's Mount Washington.

Women's basketball was more popular than men's, but in neither was dribbling part of the play. The third Boston Marathon was won with a time of 2 hours, 54 minutes, and 38 seconds.

Photocopying, magnetic sound recording, underwater remote control photography, a liquid-center golf ball called the "bouncing billy," and the automatic cannon were invented.

Writers E. B. White, Ernest Hemingway, Vladimir Nabokov, Hart Crane, Jorge Luis Borges, and Elizabeth Bowen were born, while a restaurateur in New Haven, Connecticut, created a new sandwich that involved a meat patty between two pieces of toast, to be called a hamburger. Tolstoy wrote *Resurrection*, Oscar Wilde *The Importance of Being Earnest*, and Rudyard Kipling "The White Man's Burden." Kate Chopin's *The Awakening* was, when not simply ignored, roundly denounced for its "immoral" theme of female sexuality. Haiku as a verse form was introduced to the English language.

Edward W. Nelson, who had been a government meteorologist stationed at the mouth of the Yukon River from 1877 to 1881, published his ethnological study, *The Eskimo about Bering Strait*. Roman Catholic prayer books were printed in the Central Yupik language for the benefit of Alaska's Christianized Eskimos.

Popular nature writer William J. Long published *Ways of Wood Folk*. Among Long's "observations" were birds mending their broken legs with casts of mud and feathers, porcupines curling into balls and rolling downhill to escape predators, and a fox that captured chickens by running in circles under roosts until the chickens got dizzy and fell into its jaws. These claims would get the attention of John Burroughs. Three years later, with uncharacteristic virulence, Burroughs would rail against "nature fakers" who presented slipshod observations or fantasy as scientific fact.

Governor Theodore Roosevelt of New York gave a speech in Chicago in which he said, "I wish to preach, not the doctrine of ignoble ease, but the doctrine of the strenuous life." Meanwhile, Thorstein Bunde Veblen, in *The Theory of the Leisure Class*, introduced the concept of "conspicuous consumption." Coca-Cola, then containing cocaine from the coca plant, was bottled for the first time.

Edward Kennedy "Duke" Ellington was born, Johann Strauss died, Sibelius composed his Symphony no. 1 in E Minor, and Scott Joplin's "Original Rag" and "Maple Leaf Rag" appeared in sheet music form.

Mount Rainier National Park, containing the largest single-peak glacial system in the United States, was created by an act of Congress. Congress also passed the country's first pollution-control law, providing for fines of up to $2,500 for oil spills and similar acts of nonsewage pollution. It would not, however, be enforced. Wisconsin's last wild passenger pigeon was shot.

The U.S. Army's all-black 24th Infantry company was sent to Alaska to keep the peace among people en route to the gold fields. Wyatt and Josephine Earp arrived in Nome, where the beach was home to 12,000 miners and other fortune-seekers—one-fifth of the entire Alaska population. The first copper discovery in Alaska was staked along the Copper River near Cordova after an Ahtna Native showed the location to a pair of white miners.

John Brady, territorial governor of Alaska, reported to the U.S. Secretary of the Interior, "Alaska is a land wherein men must eat bread in the sweat of their faces. The thorny apple and the devil clubs are here, and they tear and hurt. Nevertheless, the command to 'Be fruitful and multiply and replenish the earth' will be obeyed."

It stood midway between the 1837 birth of John Burroughs and the 1962 publication of Rachel Carson's *Silent Spring*.

ARTISTIC LICENSE

A *string of volcanoes* runs down the west side of Cook Inlet: pyramid-peaked, white, faces streaked with evening shadow. Redoubt to the north, Iliamna, then the island Augustine. In my time all three have rumbled to life, huffing steam and billowing out dramatic clouds of dark ash.

Iliamna was huffing then, when the *George W. Elder* passed. Burroughs, recording his first live volcano, described it as "an impressive spectacle . . . wrapped in a mantle of snow, but . . . warm at heart," Muir as "glorious Iliamna smoking and steaming distinctly at times." The artist Frederick Dellenbaugh painted the mountain with a thin line of gray drifting from its summit and with snow all the way to tideline, which surely couldn't have been accurate for the end of June, when all the lowland and foothills are not only bare of snow but green. Perhaps, though, it looked lovelier painted that way—all that stark white between sea and sky.

There were trees there, in the lowlands, as there are now— the northernmost reach of a rain forest made of spruce. On its return to these waters late in July the *Elder* apparently sailed close enough to the shoreline for the tree lovers to see that the trees were dead. Muir: "on the west side of . . . Cook Inlet considerable areas were covered with dead forest, said to have been killed by showers of ashes and cinders . . . from Iliamna; some say by ordinary forest fires."

Said by whom?

I'll go out on a limb and say the spruce trees were killed by neither volcanic fallout nor fire. Ash is not hot by the time it falls to earth and wouldn't kill trees, and the volcanoes here do not erupt in the style of Mount Saint Helens, with blasts of killing heat. Fire is not a major force in this maritime climate.

But this is true: large areas of forest on the west side of Cook Inlet are dead today, killed by an infestation of spruce bark beetles. Only recently have our forestry experts come to think that the successional force in the forests of this region is the modest bark beetle and its destructive grubbing ways. Mature forests— what tree-cutters today call "overmature"—are vulnerable to death by beetle, and then the forest starts up again.

Why I Worry

[Petroleum] comes to the surface on the main land across from Kodiak Island. Its existence has been known for years, but no attention was given to it until two years ago. Some people in Seattle have taken up the matter and have men skilled in the business making examination. The quality of the article is reported to be all that could be desired when compared with what is produced elsewhere.

REPORT BY JOHN BRADY,
territorial governor of Alaska,
to the United States Secretary
of the Interior, 1899

T*he governor's geography* was a bit off, the 1899 oil exploration not so much across from Kodiak as to its north, across from Kachemak Bay. As we slide out the bay and turn south into lower Cook Inlet, Oil Bay lies somewhere there on the far shore, fading behind us. That first site of oil drilling in Alaska is now just another wild westside crook in the coast.

In 1899, though, those men from Seattle had staked their claims, organized the Alaska Petroleum Company, done their preliminary work, and were ready to strike black gold. The historical chronology is weak, but some reports have them trying to land drilling machinery that year, without success. They must have come too early or too late in the season, must have missed the placid summer seas enjoyed by the H.A.E. In any case, over the first few years of the century they drilled several holes and succeeded in finding oil—at best, a flow of fifty barrels per day. The wells, though, had an unfortunate tendency to cave in or, when they did produce, to raise more gas and water than oil. They were abandoned after 1906.

As it happened, the first Alaska oil fields to be developed in a big way were discovered farther up Cook Inlet, where production began onshore in the late 1950s and offshore ten years later. Today, twelve offshore oil and gas platforms and 230 miles of undersea pipeline suck smaller and smaller amounts of oil. And every year the industry dumps millions of pounds of toxic pollutants into Cook Inlet waters, to be current-flushed down to this lower inlet and out into the Gulf of Alaska as though the marine system were one giant toilet bowl.

I hold fast in my memory the protesting time twenty-some years ago when this lower inlet area peaked again in the imaginations of oil companies and politicians. New leases were proposed, elaborate plans laid for everything the development of a major oil field would require. At public hearings, fishermen set king crabs before the bureaucrats and argued *oil and fish don't mix*. Local residents, myself among them, pled *why here, why us, this rich and difficult environment, the risks are too great*. The federal government, acting, so it was said, in the best interests of all Americans, let the leases.

And then, at considerable expense, the industry drilled a series of dry holes. We let out our collective breaths.

Later, what we all feared did happen, but not as any of us in our darkest dreams might have imagined. Oil from far-off Prudhoe Bay, drained by that knifeslash of a pipeline down the middle of Alaska, was loaded aboard a tanker with a drunk skipper, the tanker driven onto a reef three hundred miles from here. Day after day, wind and currents spread the oil south and west, until, weeks after the *Exxon Valdez* spill, that poisonous crude washed all the way to Cook Inlet and up against its western

shore, and then down, all around Kodiak Island, and out along
the Alaska Peninsula, through all our fish-rich waters and up
over pristine beaches.

And now, again, the federal government proposes a new oil
lease sale in these waters. Its own environmental impact state-
ment estimates the chances of a major spill at between 27 and 72
percent. The area remains as vulnerable as ever.

I stare at the western shoreline, all its nooks and crannies and
present wildness, and I think about our boat running on diesel,
the *Elder* on coal, those two dirty and exhaustible fuels. When
we have public hearings now, oil workers from up the inlet arrive in force in their pick-up trucks with out-of-state plates, their
oil company caps and cowboy boots. They always lecture us that
the building we're lucky enough to be meeting in—the school or
senior center—was built with oil dollars and we should be grateful. They like their jobs, their steady paychecks. They say they
like to catch fish from the inlet, too, and they're not worried
about pollution. They've seen worse.

We curve away south, on the same curve the *Elder* took out of
Kachemak Bay and down the inlet. The steamship's passengers
were most likely unaware of the wildcatters waiting at Oil Bay
around greasy iridescent seeps. As a group, they would have
been interested, supportive, eager for the development of another resource, another source of American wealth—just as they
had admired the bit of gold dust wrapped in newspaper shown
them by a Yakutat storekeeper and, before that, the thundering
Treadwell gold and quartz mine in Juneau. Burroughs would not
have been concerned. He would write, another day, "The fuel in
the earth will be exhausted in a thousand or more years, and its
mineral wealth, but man will find substitutes for these in the
winds, the waves, the sun's heat, and so forth."

THE BIRDING ART

Burroughs *was not* without complaint. For much of his life, and often during the Harriman expedition, he was known to find fault with "professional ornithologists." They had, he thought, strict and bloodless scientific views, and they obsessed about collecting specimens instead of bothering to observe birds in their natural settings.

At a stop early in the expedition, he took a look at "the shabby old town" of Wrangell and its totems while the ornithologists— the "real" ones, the scientists—scattered into the woods with their guns. Faithful scribe, he later recorded without comment, "Our collectors brought in a Steller's jay, a russet-backed thrush, an Oregon junco, a gray fox sparrow, a lutescent warbler, a rufous-backed chickadee with nest and eggs, and a Harris's woodpecker."

This was the age, after all, of scientific acquisitiveness, when nothing was too much or too minor to end up at the Smithsonian or in any other museum or university. More was better. Even the ordinary household birdwatcher kept in his possession not a life list of observed birds but a stack of bird skins. There were, admittedly, some practical reasons for this. Field glasses were still primitive, and it was difficult to clearly identify birds from any distance. Visual sightings weren't accepted as valid identifications. Actual birds, as well as nests and eggs, were routinely collected and studied by both scientists and hobbyists.

Dr. Elliot Coues, probably the most respected American ornithologist at the time of the Harriman expedition, had recently advised birders that they should collect "all you can get" of birds of the same kind, "say fifty or one hundred of any but the most abundant and widely diffused species." Birdskins are capital, he said, representing both monetary and scientific value, to be traded with ornithologists the world over. He didn't worry about depletion, never mind extinction. "With a few possible exceptions . . . enough birds of all kinds exist to overstock every public and private collection in the world, without sensible diminution of their numbers." He went so far as to write in his guide that, if you end up with too much "bulk" in the field and must lighten your load, "throw away according to size . . . eliminate the few large birds that would take up the space that would contain fifty or a hundred little ones. If you have a fine large bald eagle or pelican, for instance, throw it away first."

Burroughs himself, in his younger days, had carried a gun on his walks and brought down any bird that warranted a closer look. He'd kept an extensive skin collection and assembled for his wife's parlor a decorative case of fifty stuffed and mounted birds. In 1865 he wrote a correspondent, "As to shooting the birds, I think a real lover of nature will indulge no sentimentalism on the subject. Shoot them, of course, and no toying about it."

We know this: the last passenger pigeon Burroughs was ever to see was one he shot dead himself.

Advanced Learning

For *years I have wondered* about College Fjord in Prince William Sound. Why, in the middle of Alaska's coast, should there be a water body called this, and around it a whole series of glaciers named after elite eastern schools?

As we proceed through lower Cook Inlet, away from glacier country, I bury my head in Burroughs's account and find—not that I'm looking for it at the moment—my answer. First this: when the *Elder* entered the sound and nosed up to an imposing two-hundred-foot-high and four-mile-long wall of ice, "We named this the Columbia Glacier." And then, a couple of days later, when the ship pushed deep into the sound and up its farthest glacier-ribboned arm, the naming fever apparently struck again.

I set the scene in my head: the ship "coquetting with glaciers," as Burroughs puts it, all those ice rivers pouring from the mountains in ribs and blue wrinkles. The whole company on deck, in the bright sun that has them all squinting, shading their eyes from the everywhere glare of snow and ice and floating bergs. All around, thundering, echoing explosions: the glaciers shattering, dropping chunks sometimes the size of houses. The air is icy, and Burroughs is cold, hugging himself in his greatcoat.

But there—the party counts together—they see five separate glaciers all at once, and more ahead, more behind. Someone, here at the closing of an age, apparently feels the need to name these all, and so they do. The two largest, at the fjord's end, become Harvard and Yale (never mind that they'd previously been named Twin Glaciers). The smaller one attached to Harvard must be Radcliffe, and then the others lined along the west side from Radcliffe must be Smith, Bryn Mawr, Vassar, and Wellesley. The main one on the east side gets called Amherst.

What ownership is taking place here? I'm not sure I know. Harriman's invited guests come from a variety of distinguished institutions—Harvard, Yale, and Amherst among them. They are collecting all manner of Alaska specimens for their schools and museums, and now they also seem to be bagging a few pieces of landscape. But, apparently there are "right" names for glaciers, as there are "right" schools. We find here no glacier named for the University of California at Berkeley or the University of Washington in Seattle, two other institutions represented on board, or the South Dakota Experiment Station, workplace of one of the botanists.

Burroughs himself has no university affiliation, is without much formal schooling. From the start of the trip, in fact, he has felt out of place among the academics and their erudition. The zoologists and botanists, he wrote to a friend, speak Latin most of the time, and the geologists have their own incomprehensible language. "I keep mum lest I show my ignorance," he wrote, but then, in the very next sentence, "Oh, these specialists, who cannot see the flower for its petals and stamens, or the mountain for its stratification!"

I want to ask the expeditioners, do the glaciers look any different with names? More distinguished, perhaps? Better defined? Do any of you know them more particularly, for their tags of ownership?

But wait. The naming is not over yet. Later the same day, farther westward, the ship reaches the head of an inlet and a glacier called Barry. Harriman directs the captain up close, and, sure enough, a narrow passage opens between the tongue of ice and the point of land. "We shall discover a new Northwest Passage!" Harriman declares. The captain, quite naturally, hesitates; nothing beyond here exists on his chart. Harriman insists. He didn't get to where he is by being cautious. If the ship runs up on a rock—of course, the responsibility will be his own.

They slip forward. I imagine silence on the deck, as all aboard consider their radical moment, the ecstasy of setting their eyes upon a last mysterious corner of the planet. As Burroughs puts it, they are "where no ship had ever before passed." I would suffer almost any nineteenth century indignity to have been among them, to have listened to the bow cut its path into the great unknown, to the edge where the earth drops off and monsters dwell. It would feel like that to me, that off-the-chart notknowing, a last luxury of the past.

The beyond glaciers, says a zoologist, look like "the stretched skins of huge polar bears." One the group names Serpentine, for the way it winds out of the mountains. Another they call Stairway, to memorialize its peculiar blocky fall into a series of terraces.

The big glacier at the head of the fjord, some dozen marvelous miles in, must, of course, be christened Harriman. And the body of water itself, Harriman Fjord.

This, as far as I know, ends the course of the H.A.E.'s geographical naming. *Here* the fancy eastern colleges, and *here* the man who dared the unknown, a man with a grade-school education and a risk-taking genius. The two affix to the Alaska landscape in near proximity, as circumstantial as a steamship jog and as long as the ice shall last.

The Whiteness of the Way

Snow peaks. Glaciers. Breakers along the shore, and whitecapped waves, a ferociously whipped sea. Clouds. Fog. Mist. Icebergs floating past. The unsummery whiteness he met again and again—in Glacier Bay, Prince William Sound, up through the Bering Sea—benumbed and beleaguered Burroughs. Wrapped in his oversized, borrowed coat, his hat snug on his head, he complained incessantly of the cold. The glacier-fed bays were "refrigerating chests," where "the air had all been on ice, and the sunshine seemed only to make us feel its tooth the more keenly."

That ghastly whiteness. White, Melville wrote, was not so much a color as the visible absence of color, and at the same time the most concrete of all colors. Not purity, then, but something that could panic the soul. It was the whiteness of the albino whale that above all things appalled Ishmael. The possibility of a colorless, all-color atheism, a big dumb blankness.

White landscape, apparitional.

On it, in it, over it, under it—white creatures, meant to disappear into that same all or nothingness.

Burroughs and his fellow expeditioners noted this: ptarmigan above snowline still dressed all-white in June, while the ptarmigan at lower elevations had refeathered into browns that matched the ground upon which they nested. Birds of both plumages depended on not being seen, so much so that one darkly disguised brooder remained on its nest to be taken by hand.

On islands in the Bering Sea, snowy owls perched like remnant snow drifts upon a tundra carpet stitched with willows and embroidered in pink and yellow.

The snow bunting—called by the expedition's ornithologists the "hyperborean snowflake"—was found there, too, nesting in rock crevices. Chastely adorned, its plumage "as candid as a freshly opened lily," a "wraith of the snow," wearing the "ermine of kings"—all this was said of its mere and fluttering whiteness.

White gulls, too, all along the coast, and especially the northern one they called after Point Barrow, in its whiteness the sea's equal to the land's snow bunting.

Snow geese, and swans—tundra, trumpeter, whooper. All of them white.

The white-furred arctic fox. The white-skinned white whale, the beluga of Cook Inlet and the Bering Sea. Polar bears, whose white is truly the absence of color—their hairs translucent and hollow, taking in ultraviolet light at their tip ends and conveying it through those hollow shafts to black, heat-absorbing skin. It's said that the polar bear's nose is so black against its all-white world, it can be seen miles away—with binoculars, from up to seven miles. And it is said—though this may only be the stuff of legends—that the polar bear, when hunting, will hold a paw over its nose, will hide that one give-away black thing.

THINGS WE CARRY

The Elder *had its musical* instruments, its evening lectures and entertainments, its fresh milk and fine china—the high culture and convenience it carried through Alaska's wild waters.

The *M&M* has its culture, too.

The string of Christmas lights shaped like M&M's candies, strung along the front of the pilothouse. The plastic drinking glass tied with fancy knots in the window—a vase we'll fill with wildflowers as soon as we go ashore somewhere and find some. (Ken says the fishermen who come aboard to talk with him get all soft and doe-eyed at the sight of flowers.) The bread-making machine with its yeasty warm smells, the microwave, the cases of microwave popcorn, a bunch of green bananas.

The panel before the captain's chair like the console in a jet: switches, gauges, meters. All the electronics. The big pair of binoculars. Black marker on the wall: *Only one steering pump at a time!* Lists of radio channels and codes. Cup holders. A can of Bag Balm and a spring-tightened gripper for exercising hand muscles. *Chapman's Piloting, The Book of Best Loved Poems,* an old *Rolling Stone* with Drew Barrymore on the cover. A lightbulb shaded with a cutaway Coke can.

A photograph of the *M&M* at the shipyard when it was new, all shiny blue and white, a crab boat too late for the crashing crab fishery. Photos of the owner's family in Hawaii and on their gill-net boat. Drawings by the owner's younger son of both the gillnetter and the *M&M*—a child's cartoonish drawings with blue wavy waves and stick-figured humans but with amazingly accurate detail to the boats themselves—near-perfect proportions, exact rigging, even the below-water appearance of keels and props, anchor and anchorline, schools of salmon swimming past.

A yellow sticky note: *Only the truly mediocre are always at their best.*

A Hagar the Horrible comic strip. Hagar to Helga: "My crew has demanded a raise so I have to go make the final arrangements." Helga: "What final arrangements?" Hagar: "Firing them and getting a new crew."

In the galley, hooks and bungee cords holding every door, cabinet, and drawer shut. Pot holders with cows on them.

Over the bed, a gun rack with one rifle and one silver, purple, and metallic blue cheerleader's pom-pom.

Downstairs, nailed over a doorway, the open mouth of a shark, from its lips to its gills. Double rows of finely pointed teeth. The dorsal fin, like a hunk of old leather, dangles from a string alongside.

Ear protectors to wear in the engine room. The refrigerated sea water system for keeping fish fresh in the hold.

Evening, and three of us in the galley laugh our way through Woody Allen's *Bullets over Broadway*. The younger crew members, in their quarters "before the mast," watch a different video, something with *blood* in the title.

It feels to me complete, this world floating within steel, on water, across space. We are its citizens, we six with our pocket knives and enthusiasms. The boat has its storied life; we have our passage. Another laugh, a shake of head, the view from the wheelhouse window, more than enough orange survival suits stuffed behind seats.

Religion

Biographer *Perry Westbrook* wrote that Burroughs at age seven or eight was "strangely thrilled" by a glimpse of a black-throated blue warbler in woods near his Catskills home.

Why is this so strange? To whom is it strange? It strikes me as strange to think that such behavior is unusual or odd. A black-throated blue warbler—this small bird with a white spot on its wing—may not be a particularly showy bird, but who needs showy?

When I was young I used to play, with other children and often by myself, in "empty" fields at the end of our city block. It was my kingdom—that country of granite-block cellarhole, swooshy pines, a lone beech tree carved with certainly ancient hieroglyphics—as complete a world as that mapped out in the book of Pooh. I climbed into the trees during windstorms, startled rabbits in their pathways, held grasshoppers in my hands until they squirted "tobacco juice." I built hay-sided, open-roofed forts where I lay in prickly comfort to watch parades of clouds.

I know I was often thrilled by what I found in my kingdom. One day, though, stands out for me with a particular color clarity. It was there in the field I saw a slice of blue sky on the wing. Just a glimpse, all blue, and gone. A shot from the grass, a pulsing past milkweed stalks, a disappearance into trees. It was motion, light, one unexpected moment of flashing beauty, and all to myself. I searched the trees and the fields and sky, wearing the winged vision in my eye, but I couldn't find my bird again.

I ran all the way home and, breathless, told my mother what I had seen. I was a child, seven or eight; I did not have the words. Perhaps I seemed "strangely" excited. I said I'd seen a blue bird. Together, my mother and I looked it up in a field guide. An indigo bunting. *Indigo bunting.* I had never heard of such a thing, not the word *indigo*, not the word *bunting.* The words themselves were exquisite. Of course, there was something satisfying about putting a name to what I had seen, to having it authenticated. It existed, the bird. But it was not just a bird. What I had seen was something unnameable, ineffable, a recognition of something else—beauty, yes, but also mystery, a rightness in the world.

Years later, when I read the works of James Joyce and Hermann Hesse, I learned the word and concept of *epiphany*. I remembered my indigo bunting. I had a name for that peak, private moment of recognition.

There was another time, about the same age, when one early morning I lay under a berry-bearing tree flocked with birds. I believe the tree was a mulberry and the birds waxwings, but what was significant was the way the low branches bent down around me, enclosing me as in a hidden chamber, and the oblivious birds, with crayon-colored red and yellow feathers, fluttered above me, their stout bills cracking open and shut. I watched how their feet closed around thin branches, and the way the breeze parted the feathers of their breasts. I saw the light in their eyes. With them, I plucked and devoured the juicy berries. We weren't so different, those birds and I. I knew that with a child's communicant clarity, with a shivery thrill.

Children have these moments, hiding in bushes and climbing among branches. Adults neither hide nor climb, and they never expect to be surprised. They look elsewhere for their thrills.

I have to love a sixty-two-year-old man who, when he stepped ashore in Alaska, spread himself on a mossy bed beneath spruce trees and waited two hours for the singer of an unfamiliar song to reveal itself. And who then wrote his discovered thrush, "our robin in a holiday suit," a long love poem full of corny rhyme and exclamation point.

O Varied Thrush! O Robin strange!
Behold my mute surprise.
Thy form and flight I long have known,
But not this new disguise.

Rainbow

whisper in my ear. My eyes snap open. I'm back with the tender's diesel grind, the roll of the sea. Ken, on the edge of the bunk, is already rising, beckoning.

I slip on my sweatpants and follow through the galley to the wheelhouse. There. On our port side a full rainbow scoops the pale sky from sea to sea. The bands are wide and deeply hued, colors purer than paint.

Barefoot, I escape past windows to unstoppered air and stand cross-armed at the rail. The perfectly formed arch rises as out of two blow holes in the sea, holding up the sky. Above it circles a second, less brilliant bow, like a tentfly to a tent. Centered by both, perfectly framed across the water, lie the Barren Islands. Although the near-midnight sun has set on us, its glancing rays, aimed sidearm from over our northwest horizon, have lit the islands like fire. The treeless crags glow impossibly green, more Irish than all of Ireland.

I imagine aiming the boat toward the rainbow's western end. Surely, what I see is so substantial, it does drop just there into the sea, as solid as a stone column. It must be reachable, touchable. Any reasonable person would want to go out of her way to be showered upon by gold and majestic violet, to grab fistfuls of dazzling color. To know the physics, the inevitable disappointment of chasing rainbows, doesn't change my desire.

I remember that the Makah Indians of what is now Washington State thought that rainbows had claws at either end to seize the unwary or overzealous. Perhaps that was a lesson of their own history: a rainbow chased, the chaser lost. I picture claws like giant crab claws, grounding the rainbow to the sea floor, now scuttling sideways, out of reach.

What was it Burroughs wrote of rainbows? A rainbow is "one of the most lovely and wonderful things in nature, and yet it serves no purpose in nature; it has no use."

What a curious concept, I think—dividing the natural world and its phenomena into the useful and not-useful. To a farmer, rain is surely useful, and earthworms, seeds, even rocks that will make walls and—eventually—new soil. The red color of berries is useful, because birds can see red, can thus find food, and, in a very useful reciprocal arrangement, will carry off the berry seeds to sow elsewhere. Burroughs was known to despise woodchucks, and yet I'm sure he would find them useful for their contributions to the web of life, no less than for keeping his trigger finger in shape. And I am just as sure he would find the tree that falls and rots in the woods to be useful. In fact, as I think of this, it comes to me that Burroughs, for his time, undoubtedly had a far broader understanding of the interdependence of all things than the average man or woman. Perhaps not as broad as Muir, who was much clearer in his articulation, but still—he understood ecology before there was a word for it.

Useless rainbow! As useless as ripples on water, as streaky pink clouds, as the sound of rain. Who can live without these?

West

We are moving mostly south, into Shelikof Strait, but it feels like *west, away, out* along the Alaska Peninsula. The farther we go, the more expansive I feel, as though the crack that lights the world is opening wider and wider.

I try to sympathize with Burroughs's reluctance to come this way at all, west all the way across the country, and north to Alaska, and out this way, out, out, so far west finally they crossed over into the next day. Even as he was leaving New York on the Harriman train and passed his place on the Hudson, where he thought he saw his wife waving her apron from the summerhouse, he feared he had made a terrible mistake. He wanted to be home, not so much with his wife, who had never shared his passions for nature or literature, but with his grapes and his familiar birds, trees, the hills and creeks and even rocks he knew by name.

I try to sympathize, but I don't succeed very well. I never had such a problem. The years I was growing up in New Hampshire, I always faced west. I wanted to be the little girl in the big woods and on the prairie and as far as I could get. I took to heart the words of Thoreau—my fellow easterner but so, somehow, an authority—that in the West a person's soul would have room enough to expand and not "rust in a corner." When I reached college and my first winter break, classmates and I drove straight from Massachusetts to the Colorado mountains, crossing the

Mississippi at St. Louis with the sun coming up behind us. The Rockies, rising from the plains, filled me with genuine awe: their sheer size, all those sharp new edges and points. Distances, wildness, the light shining through aspens, a people who didn't much care where anyone else came from and tended to have dirt under their nails—these drew me. West, and west. I landed in Alaska and was soon as fiercely at home there as Burroughs ever was along the Hudson.

I know, of course, the accepted wisdom that westering folk are running away from something or—and it might be the same thing—making a fresh start, taking a new hold on life. The truth of this is everywhere in Alaska, from the early goldminers and settlers to today's end-of-the-roaders making one more try at whatever it is they think will bring them success, or happiness, or love. But still, there's the *land*, and it offers more than what can be ripped out for someone's profit.

Burroughs readily admitted that he had trouble with Alaska's scale, its massiveness, unruliness, wildness. Alaska's nature was not, for the most part, the nature he was accustomed to back home, the "mother nature" within whose lap he could take comfort. He let the grand-scale adventuring go to other members of the party, and kept practicing with his eye, trying to find his range.

Small Things

I n *Alaska*, this is our Burroughs, without apology:

He is in awe of the sparrow's blazing yellow crown between black borders.

His passion lies with the elfin pink flowers of the moss campion that blooms at the edge of snow.

He delights in discovering a tiny titmouse nest in a hole along a stream bank, counts six dark-brown eggs, and notes with evident satisfaction the pussy willows just furring nearby.

He is extremely pleased with the red-vested bumblebees that sip at the lupine blossoms.

When in Sitka, he is so taken with the water ouzel's song and the way it matches the liquid bubbling of the currents around it that he gives it more words in his narrative than his tour of the town, the Indian settlement, the museum, the Russian church, and the nearby hot springs combined.

No, he cannot adjust his eye. Burroughs simply is not a man of mountains and oceans, of furrowed blue glaciers and tangled forests. Leave all that, leave the sublime, to the "other John." Burroughs is a man at home in farm fields and logged-over woods, with his boyhood bobolink and a fascination with the habits of chipmunks.

Once I heard an interview with the naturalist E. O. Wilson, in which he said he had become an entomologist because his eyesight was so poor. He couldn't see anything too far away, and so he paid attention to those things he could get close to, the closer the better, preferably with a magnifying glass. He came to know insects with absolute precision, but he never stopped with carapaces and antennae or the production of ant eggs. What he knew about insects he also knew about the delicacy and re-silience of the world and the connectedness of all things in it.

This was Burroughs's capacity, too. His eyesight was, to my knowledge, fine enough, but his vision was close, trained to the familiar detail. He surely agreed with his friend Whitman, that

> ... *a leaf of grass is no less than the journey-*
> *work of the stars,*
> *And the pismire is equally perfect, and a grain of*
> *sand, and the egg of the wren,*
>
> *...*
> *And a mouse is miracle enough to stagger sextillions*
> *of infidels* ...

Anyone might be awed by the extravagance in nature, but to find the same wonder in the small things, even and especially amid the grand—*that* is a rare talent.

WEATHER REPORT

Kadiak, I think, won a place in the hearts of all of us. Our spirits
probably touched the highest point here. If we had other days that
were epic, these days were lyric. To me they were certainly more
exquisite and thrilling than any before or after.

> BURROUGHS,
> "Narrative of the Expedition"

All night we slip down Shelikof Strait, sleighriding the
tide. By morning, gray skies lower to distant gray
coastlines on both sides. We are just past Kukak Bay on the
Alaska Peninsula, where the H.A.E. put ashore a group of
botanizers for several days. On the Kodiak Island side, we are ap-
proaching Uyak Bay, where the expedition next stopped to leave
off a bear hunting party. The botanists would be successful, the
hunters not, though Burroughs will report that the hunters "had
the comfort" of their tramps and a "superb" waterfall.

Kodiak, or Kadiak—the Native word, as the Russians heard
it, for *island*. Among all the islands of today's United States, only
Hawaii has more land mass. Kodiak's coastline is anything but
circular; it stretches like putty around long bays and fjords, into
peninsulas, hooks, and bights, past crumbed-off small, smaller,
and single-rock islands. And raw. The land is raw. The North
American coniferous forest stops here, on Kodiak, with the Sitka
spruce that darken the north end before giving way to grass-
lands.

The town of Kodiak, facing the wide-open Gulf of Alaska, lies all the way around the island from where we pass. There, Burroughs reveled in warm sun and pastoral splendor, an end to overshadowing forest and chilly glaciers.

"I shall babble continually of green fields," he wrote. And so he did, exclaiming again and again of verdant hills, green loveliness, smooth vistas, rural sweetness, vales of birdsong and flower-strewn slopes, the village strung with grassy lanes and ruminating cows, an emerald billow of a mountain, a vast green solitude stretching away in three directions.

I see him making his way around the village, past the chiming Russian church with its blue dome, into the store to ask for eggs, then to a cottage near the beach where he meets a large Russian woman with fresh eggs to sell. The currant bushes in her garden are in full flower, the potato plants a foot high. The sun beats down, children play beside the door, the varied thrush and the winter wren sing out. The ground underfoot is solid and smells of green warmth and homey cows. I think Burroughs breaks off a stem of grass and chews it as he walks yet another enticing trail, following the chatter of a creek, meeting boys with strings of trout.

He could live in such a place, he says; he could repeat forever its exquisite days. Our man about Kodiak is in love with the blend of wild and domestic, the arcadian seclusion of it all.

From our pilothouse, I stare at the island. Gray. Gray sky, gray water, a dark, misty, mountainous gray land. The wind blows fine needlepoints of rain against the window.

Burroughs was a keenly observant naturalist, but somehow in Kodiak he forgot to wonder whence came the green profusion he so adored, and all those rills of lovely trout-stream water. He asked about temperatures, and was pleased with what he learned, but he apparently neglected, in his few days in paradise, to ask about rain.

Mountaintop

Is this the wilderness—these greensward hills,
These wastes of lupin, wildflowers and of rose,
These slopes of heather, by the mountain rills,
O'erhung by skies of gold through day's slow close,
Where one long lotus-dream obscures all human woes.

Charles A. Keeler,
"Fourth of July Ode,"
written for the H.A.E. in Kodiak

I n *Kodiak, Burroughs* climbed the mountain behind town. From the *Elder* it had looked like a smooth meadow walk, but our man found himself wading through knee-high grasses, parting gates of fern fronds, squeezing through mazes of tangled alders. His hands filled with blue and purple flowers: lupine, wild geranium, bluebell, Jacob's ladder, iris, the Kamchatka rhododendron. Near the summit, he came upon patches of fragile forget-me-nots—"a handful of it looked like something just caught out of the sky above." There, too, he found a pale yellow lady's slipper striped with maroon. I imagine him in all this splendor, hands overflowing with beauty, exalting.

It is the scale thing again, I suppose, or perhaps Burroughs's vision is less than perfect—all that reading and writing *could* have left him nearsighted. In any case, his narrative goes on at length about the mountain's flowers and birdsongs but says not one word about the view from the top. This I must fill in for him: the village at his feet, so nearly under the mountain that he might toss petals into the air and watch them float down to the church-yard. Any direction he might turn, more hills and mountains and valleys, all of them green as green, only dimming with distance, one shade at a time. The coast jigsawed into sharp points and circling bays, water touching sky, a horizon giving shape to a world.

A year ago I stood atop that same Pillar Mountain, not in summer but at the end of March, on gray tundra and scree, avoiding the hollows filled with rotting snow. The day was clear, the view wide. My friend pointed out landmarks: Spruce Island, Marmot Bay, Afognak Island in the distance, Chiniak Point, the three bays of Womens, Middle, and Kalsin, Woody and Long Islands.

We had not hiked up Pillar Mountain but, like modern peo-ple in a hurry, had driven the road to its top. The road ended at a scrap heap of old radar dishes and antenna towers.

That day, earlier, the mountain had been the site of the an-nual golf classic, a tournament that starts at the mountain's base and ends at the one hole on top. The course up the adjacent val-ley was marked with surveyor's tape tied to alders and stunted spruce, and a snow path trampled by many feet.

I looked down on streets and roofs and bulging boat harbors, on the bridge to Near Island that has brought that island that much nearer. I picked out the new museum of Native culture, and the old building where Russian artifacts are displayed—the one built nearly two hundred years ago as a warehouse for the Russian-America Company's extravagance of sea otter pelts and the one, I assume, where Burroughs spotted a song sparrow singing from a weather vane. In the boatyard, blue tarps and shiny aluminum decks shot back the light, and I knew that among them, too small to see, fishermen were mending seines, fiberglassing hulls, changing over their electronics. Across from the boatyard rested a fleet of identical white boats owned by the Moonies, a religious enterprise that each summer hosts hundreds of sport-fishing clients from Korea. Earlier, a few blocks off, in the business section of town, I had driven past a carwash called Buggy Banya, after the name of the traditional Russian steambath, and a natural food store called Cactus Flats, a world away from any desert. And there, just outside the main boat harbor, I knew that people from the local Fish and Game office were still trying to capture a sea lion being strangled by a rope accidentally looped around its neck.

In the big view, though, the entire town, with all its human commerce, was still only a dot on the landscape, a pencil point on the canvas that stretched from sea to sea. There is room there, even today, for a man or woman to tramp out from town, up one green valley and down another. Most of the island in my sight lay within a wildlife refuge, home to, among others, some twenty-five hundred brown bears.

My friend and I stood on Pillar Mountain, and she began to tell me that one time, late in summer, she'd met a man who was frantically searching the slopes for a rare orchid he had heard grew there. She said it was long past the time for flowers and she didn't understand why someone who cared about such a plant wouldn't know that.

I looked at the cold hard ground. In three months it would be carpeted with Burroughs's splendid flowers. The wild geraniums. The purple rhododendron blossoms huge against the tundra lichens. Those sky-blue forget-me-nots. And I hope, still here, that delicate yellow lady's slipper for which he would have gotten down on his knees.

HARRIMAN'S BEAR

No bears, no bears, O Lord! No bears shot!
What have thy servants done?

MUIR JOURNAL

T*he bear was the genesis* of it all. When Harriman's doctor told him to take a relaxing vacation, his first idea was a hunting trip. He wanted to kill a big, trophy brown bear. Alaska had the biggest bears, and so his destination was set.

The H.A.E. searched for bears all along the panhandle coast, teased by the occasional paw print but never spotting an actual beast. Prince William Sound, Cook Inlet, the Alaska Peninsula—they looked everywhere for bear. And found none. Kodiak, various old hands along the way reported, was the ultimate bear location, home to the largest brown bears in all Alaska and the world. Harriman set his determined eye there.

While Burroughs picked wildflowers and visited with old Russian women, the hunting party, led by a local guide, boated down the coast and hiked inland. Finally, on a sunny hillside, the railroad man shot his bear. His photograph of the dead animal (caption: *The Bear*) shows the bear's head in profile against tundra, one blond front shoulder, one thick upper leg. The mouth is open: light-colored gums, a white canine tooth, a tip of tongue, and—it looks like—some matted, saliva-sopped grasses.

It was late the next day, July 4, when the bearskin was brought back to the *Elder* by rowing boat. The kill had been celebrated for a full day already, the holiday itself filled with the firing of cannons, boat races, recitations of patriotic poetry. Burroughs's narrative is silent on the dead bear's delivery, but I take the liberty of playing out this bit of theater—this last entrance on the stage, one more pageant for the festivities:

Cheers, as the boat with the trophy pulls up alongside. All the Harriman partiers crowd the rail. Burroughs, as curious as the rest, looks first to the men in the boat—his friend the bird artist, and the taxidermist. The men, under their caps, are burnt by the sun, are weary, are smiling. They'd left at three in the morning in order to get the bear back on the holiday, a fitting end to a day that celebrated America and all the dreams of all the greatnesses Americans might achieve. In this country, anyone might amass a fortune, become famous, succeed in whatever task he set for himself. Wasn't this bear, this trip, this gathering in this place proof enough?

Burroughs looks then into the bottom of the boat: the bunched hide, the floppy head, the black pads of an upturned foot—these baggy outer clothes stripped from the real and whole creature. For a second, perhaps, he thinks of the rest of the bear, abandoned on a hillside, a mound of flesh left under the sun to burn and rot and be pecked apart by birds and chomped by fox, to finally disengage into a loose pile of bones.

The one bear, and the second one, too—this year's cub, smaller than the smallest Harriman child. Once the mother was dead, there was nothing to do about the bleating young one except kill it too, this pitiable creature. Something—blood?—is smeared across the small one's fur, and flies are gathered at its lips. The boat carries with it the hot, meaty smell of death.

I construct more details: the trophies hoisted onto the deck, the fanfare of stretching out the hides, stringing them up in the rigging. Everyone wants to stroke the fur, to lift the heavy paws and examine the straight, daggerlike claws. The women coo over the cub as though it were a living pet. The children touch the teeth and jerk back their hands. Someone makes a joke about dental hygiene.

"A real big genuine Kodiak Bear," Merriam declares. Well, it is real, and it is genuine, and it *is* big, compared to, say, the Sitka deer that had been shot earlier in the trip. It's just that this particular specimen of Kodiak bear is small compared to the average, very small compared to anything that might be considered a true trophy. It is a puny young bear with patchy fur, and punier still against all that has been spoken these last two days, all the talk of great bear hunts, a great bear, a great hunt, Harriman's lifetime dream.

Burroughs can see this, as he sees that Muir has separated himself from the rest, is speaking with an old man from the village, both of them saying how much they like exploring the mountains by themselves. Burroughs has seen what he must, and now he will write his one paragraph about Harriman's good luck, and that the bear was eating grass like a cow when first spotted, and that it was, though "large," below the size of the traditional Kodiak bear.

When the trip nears its end, after hunting parties again and again seek and fail to find additional bears, Burroughs will add, "Nothing is plainer than that one cannot go to Alaska, or probably to any other country, and say, 'Come, now, we will kill a bear,' and kill it, except as a rare streak of luck."

Fox Dream

ML. *Washburn has a vision.*

He has seen the future of Alaska, and it is populated with legions of small blue-phased arctic foxes and the prosperous men who will manage their care and then strip them of their world-fashion fur.

Washburn, who joined the Harriman expedition back in Prince William Sound, has reached his destination. At Long Island, near Kodiak, he tours members of the expedition around his company's fox farm, lecturing them so righteously about good husbandry, opportunity, and profits, that even a railroad tycoon must be impressed.

Burroughs, along for the tour, hears the case made:

Since Alaska's sea otters and fur seals have been hunted practically into oblivion, and since populations of furbearers all over the world are in sharp decline, here's a chance both to save animals from extinction and to make up for the fur shortage.

The physical situation is perfect. Alaska has hundreds of islands that are, in Washburn's words, "unoccupied" and "now useless." Stock them with a few foxes, and—*voilà*—you have a new industry. Farmers simply feed the foxes as necessary and then harvest them in a controlled manner when the fur is in peak condition. Food for the animals is essentially free: salmon and seal meat, with a little cornmeal mixed in. On some islands the foxes don't even need to be fed except during the worst winter weather, because there are so many birds for their taking.

The pelts sell for, on average, twenty dollars apiece.

The foxes breed quickly.

There is enormous wealth to be made.

Thirty islands have thus far been turned to such farms or ranches, since a modest beginning fifteen years earlier. It is Washburn's company, the Semidi Propagating Company, that sells foxes from one farm, one island, to another.

Alaska's Natives, who now have too few wild furbearers to hunt and diminished runs of salmon, can find excellent employment in this new industry.

Once fox farming is perfected, the same domestication can be extended to other furbearers, saving them all from extinction.

And then, from Alaska, the day will surely come when farmers all across the northern states will devote small wire-fenced enclosures to this same industry and reap from it a far greater return than from all their other livestock combined.

Burroughs takes this all in, trying to sort truth from evangelical chaff. He studies the scene around him: the pleasantly wooded island with grassy openings, the keeper's cottage just up from the beach. The cottage and outbuildings are engulfed by grasses and fireweed stalks, the first fireweed petals just released to flame. There are, so Washburn says, a thousand fox on the island, but they seem to be more wild than domestic, off cavorting in the woods and fields. The only ones the visitors have seen were a couple of whiskered faces that peered at them from around the corner of the barn.

Washburn points out the box traps used to capture the animals in the prime-fur season from November to January. He describes how the best specimens are released for breeding purposes—marked by clipping their tail fur—and the others "harvested," like so many ears of corn, or like the grapes Burroughs raises back home. Washburn refers to the fox as "Sir Reynard."

A little apart from the others, Burroughs sidles into the edge of woods. From deeper within sounds a persistent, many-voiced yapping. He thinks he catches sight of dark fur—maybe the tip of a flagging tail, turning down a path. At home the foxes that come around are of the furtive red variety, and he seldom sees any more of them than the straight lines of their dainty tracks in snow. For a moment, he imagines himself living on this Long Island, tending to a garden plot as well as to foxes, and having this whole new book of Nature to explore, to get to know with more than a passing nod. The farmer's situation here really seems quite comfortable, the environment entirely benign.

But the others are moving on, and he must hurry to join them.

They all pause back at the beach, where a succession of long, high racks are hung with split salmon that, once dried, will be fed to the foxes all winter. The salmon has a fishy, slightly spoiled odor to it, and a tarnished red color. Tuxedoed magpies are all about, swooping from trees and racks, tearing at the fish with ravenlike cupidity. They have a lovely way of fluttering down with wing and tail feathers spread like oriental fans, the white of their shoulders and wingtips flashing. *Yak-yak-yak*, the magpies shout.

Now Washburn is again saying something about world fur markets and the excellent quality of these furs, and that the potential, really, is absolutely *unlimited*.

Burroughs watches one magpie in particular, decides to see how close he can come to it. He walks loosely, evenly, toward it, without sudden movements. The bird, astride its pole, eyes him fearlessly and pecks at more fish. He walks closer, closer, and the bird only keeps watch; its jet-black eye is a bead, but one with a shine a thousand generations deep. The bird turns on the pole, and the blue gloss of its back shimmers in the light, so smooth, so cool, like a cape of finely napped velvet. Burroughs walks right up to the magpie, and then stands absolutely still. He is still standing there, caught between the dark eye and the royal blue sheen, long after the others have moved on to view the skinning shed.

PART II

WAKE RIDERS

I *stand on the tender's bow*, leaning forward, hands gripped tightly to the rail. At ten knots, we're plowing through the sea, parting it with a sound like crushing paper. Below me, out of the vast blue, porpoises slice in from left and right. They catch the push of the bow's wake, and then shoot across one another and cut back into the wake again. Now their dorsaled backs split the surface with steam-kettle hisses, now their white-patched sides flash.

It must be like surfing, I think—only below the surface. There's no question these animals are at play, with the boat and one another. They're along for the ride.

These are Dall porpoises. Among the fastest of any marine mammals, they are known for appearing out of nowhere to ride the bow wakes of ships. Their black-and-white markings are distinctive, the white edges to their dorsals and flukes like matched accessories to their patches. As a rule, they get little attention from humans; they don't leap from the sea like other porpoises, and they don't survive well in captivity. No one knows how many there are. They accompany ships in a range that runs from the Bering Sea all the way down to the northern California coast. They have sharp teeth and eat a diet of squid and mollusks as well as fish.

The Harriman expedition was William Dall's fourteenth trip to Alaska. On his first, in 1865, he'd led an expedition to study the feasibility of an intercontinental telegraph line. Later he had authored *Alaska and Its Resources*, still in 1899 respected as a major—perhaps *the* major—sourcebook for the territory.

Dall porpoise, Dall sheep, Dall limpet—and several other species—all named for the man who first scientifically identified them.

Dall was the H.A.E.'s general lecturer about Alaska's discovery, exploration, and resources. He was frank about the Russian exploitations, critical of the thirty-year American occupation he considered one of anarchy, neglect, and indifference. I hear him proclaim what he later set in print for Harriman: "A history of conditions in Alaska from 1867 to 1897 is yet to be written, and when written few Americans will be able to read it without indignation."

The small whales cross and recross our bow, all quicksilver motion and light, two of them, or five or six, or eight in view at a time. Their pace seems effortless. Again and again, they appear to come within inches, *an* inch, of crashing into the bow, of colliding with one another. But they never do. They are speed and precision artists.

Then, just when I think I know something about the way Dall porpoises behave in a bow wake, five of them together torpedo up from the depths at one side of the bow. Before they hit the surface, they peel off from one another, arching back in black-and-white symmetry, opening the formation like a flower, like fireworks, like aerobatic fighter jets or synchronized swimmers in perfectly choreographed performance. They spray through the surface, then circle under, and they're back to riding with the bow.

DEVIL FISH

*Many whales were seen blowing, their glistening backs emerging
from the water, turning slowly like the periphery of a huge wheel.*

BURROUGHS,
"Narrative of the Expedition"

How right he has it. A surfacing whale cuts the water
with such an arc—turning, turning, turning—that
you think the animal must be circular.

We meet our first large whales in Shelikof Strait. They are off
to starboard, but so distant all we can see is their blowing. They
must be gray whales, which tend to hug the coastlines in their
long migrations between Baja California and the Bering and
Chukchi Seas, and which have shorter, more dispersed blows
than the humpbacks. These vaporous breaths hang over the
pitching sea before drifting like mist in the wind. No sooner has
one dissipated than another shoots up and then two together, es-
tablishing not a pattern, exactly, but a sort of pinball rhythm.

I think of my child-self, how I was New England–raised on
sea chanteys, the countless times I shouted *thar she blows!* I was
crazy for those three words, for the archaic and windy sound of
them and for the excitement they conveyed, an excitement that
was somehow caught up in my mind with peg-legged pirates and
treasure chests. I hadn't the dimmest idea when I shaded my eyes
with my open hand and pretended to peer into the distance, that
what I meant to see was anything like what is beside me now. It

was not until I read *Moby-Dick* as an adult that I understood the first thing about sighting whales at sea, and then it came like a revelation—*Oh, blowing, a whale.*

Devil fish, the Yankee whalers called them. More than other whales, the grays were protective of their young, and were said to attack boats that got between mothers and calves. Even so, these slow-moving whales were no match for the whalers and their harpoons. The Atlantic grays were wiped out entirely by the 1700s, the Pacifics pursued to the brink in the next century.

I picture the Harriman women, traipsing out onto glaciers in their long skirts. Of what were their corset stays and skirt hoops made, and their umbrella handles? For centuries there had been nothing else in the world so flexible and resilient as baleen, those fringed and horny plates from the great whales' mouths. Only when whales became too rare and baleen too expensive were merchandisers forced to invent substitutes (eventually plastics).

By the time of the H.A.E., too few gray whales were left anywhere in the world to hunt, and the last hungry whalers were concentrating on arctic bowheads. The whales Burroughs saw, in southeast Alaska, were likely humpbacks, and others of the party watched killer whales in Prince William Sound, but the H.A.E. didn't encounter the whaling fleet itself until Port Clarence, far in the northwest. There, the few ships still working waited for the icepack to open enough for them to reach the bowhead's summering grounds and last refuge, discovered just ten years before.

My breath holds in my chest as I scan for the next explosive sign of whale. I breathe out, slowly. In. Out. We're not simply a lone boat floating on a rolling blue surface. We're part of a complex, largely hidden vertical life inhabited by whales the size of ships, single-celled diatoms, crab larvae and barndoor halibut, swarming schools of pollock, salmon sniffing their way home, sea stars, sea lettuce, sea lions, seagulls, sooty shearwaters, and squid. We're traveling together, all of us—dependent, whether we know it or not, on one another and the good health of our ocean home. The puffin dives after the zooplankton, and the otter relies on the kelp for anchor, and the incoming salmon feed on the out-going salmon fry as the salmon fry eat of the old dead spawners.

Our boat is off to the fishing grounds no less than the old whalers were off to the whaling grounds, though we know the difference between whales and fish and the need for both to prosper. Beside us, on their parallel course, the fearless barnacle-ridden whales are off to their summer grazing in fields of amphipods, in this season of fattening.

Among the Whalers

Oh, we are the Innuit people,
Content in our northern home;
While the kayak's prow cuts the curling brow
Of the breaker's snowy foam.
The merry Innuit people,
Of the cold, grey Arctic sea,
Where the breaching whale, the aurora pale
And the snow-white foxes be.

William H. Dall,
"The Song of the Innuit"

At *Port Clarence*, the *Elder*'s farthest north stop, a dozen whaling ships sat on their anchors. Eskimos, two hundred or more, were camped along the beach to trade with the whalers and, now, with the H.A.E. From their walrus-hide umiaks, overflowing with children and dogs, they shouted up at the *Elder* like carnival barkers and waved their wares. Four dollars for a caribou hide, three for a pair of walrus tusks, two for sealskin boots with thick walrus-hide soles, just one dollar for a bit of carved ivory.

Burroughs observed, especially, the women with their open faces and white smiles, and determined that some of the younger ones were "fairly good-looking, and their fur hoods and fur cloaks became them well." As though he were watching yet another new bird, he noted the precise way one woman tucked her baby under

her parka from the rear, a bending and thrusting and wiggling series of motions that finally positioned the child's face over the
shoulder; he compared the movement of the baby under the
clothing to a big morsel making its way down a chicken's gullet.

Some of the party went ashore to visit the Eskimo camp,
which stretched for a mile along the shore. Our friend John did
not. He chose instead to accompany the ship to its watering at a
stream mouth. This excursion gave him several hours to explore
the boggy tundra, where he was pleased to discover the pink-
petaled spring beauty, masses of shooting stars, saxifrages, and
fat pollinating bees. He flushed a savannah sparrow and chased
after some warbling golden plovers—"the voice of the tundra,
soft, alluring, plaintive, beautiful." He witnessed a thrush hover
in the air, something he had never seen before, and he decided
that, since there were no trees on the tundra, the bird was perch-
ing instead upon the air.

Just across the peninsula, fifty miles distant, Nome was ex-
ploding with gold fever, a development those aboard the *Elder*
first learned of here. Between the miners and the whalers, the
Native people of this place were in a deadly squeeze.

Burroughs's fellow traveler George Bird Grinnell noted with
disgust that the silver with which the H.A.E. expeditioners
bought their far-north souvenirs would soon find its way aboard
the whalers, traded for cheap whiskey. He would later write
about the Eskimo's "gloomy" future: "There is an inevitable
conflict between civilization and savagery, and wherever the two
touch each other, the weaker people must be destroyed."

If only Burroughs had applied his attentive powers that day
less to the hovering thrush and more to the people in their skin
clothes, with their ivory-tipped harpoons, their grass baskets,
their dried fish . . .

I will him where he will not go, to the Eskimo camp, where
he will take in every detail, every simple act. There, with Grin-

nell, he will notice behind several of the tents sticks supporting small carved figures painted black and white. One seems to represent a bear, another a bird. One includes human figures on either side of concentric circles. A man standing by this last is asked, with gestures, *what does it mean?* His English is rudimentary, but he says something that sounds like "all same sun."

Burroughs, so atuned to the individual songs of birds, to the variations in sparrow trills as he traveled farther and farther from home, should have been able to hear this, should have thought about what it could mean.

All same sun. *All same sun.*

He, who loved Whitman so, might have found a brother philosopher.

"ESKIMO CHILDREN

FROM BERING STRAITS"

Burroughs made little mention of the Eskimo people the H.A.E. met at Port Clarence or of their corresponding kind across the strait at Plover Bay, Siberia. Photographer Curtis, though he snapped here and there, had yet to make Native people the passionate subject of his art.

I have, instead, another bit of documentation. From an 1899 report of Sheldon Jackson, U.S. general agent of education in Alaska, I keep a photocopy of a page showing six children, five girls and a boy. They are Eskimos of northwest Alaska, whom we call today by the name they use for themselves—Inupiat, the "real people." (Inuit or Innuit, also sometimes used, applies broadly to Arctic peoples living from Greenland to Alaska.)

Two photographs—one above the other. Both were shot at the Carlisle Indian School in Pennsylvania, to which the children were taken by Sheldon Jackson.

The *before* photo, taken in 1897, has the children carefully, symmetrically arranged before a white sheet, the tall boy in the center, the two tallest girls on either side of him, the next two tallest beside them, the smallest seated on a low stool in front. They're wearing wrinkled and perhaps not entirely clean cloth parkas that hang, on the girls, to their knees. The parkas are hooded, and the hoods are up, circling their faces like oval frames. Four of the hoods are trimmed with the deep light-tipped fur of winter wolves. Except for the smallest girl with narrow, Asian eyes, who sits with hands folded against her knees, the children stand erect, sealskin-booted feet together, arms at their sides, staring into the camera. Their faces are without expression.

In the second, lower photo, one year later, the children are enlarged, closer to the camera, four seated in a row and cut off at the knees, two standing behind. The girls are wearing dark Victorian-style dresses with high collars, long sleeves, pleated fronts, and shoulder appendages that stick out like bat wings. The marks of a steam iron are visible. Their hair is severely parted and tied back. Their hands are politely folded in their laps. The boy sits spread-legged in an ornate spindle chair, holding a hat in his lap. His hair, too, is parted in the middle, and his jacket is buttoned tightly with four brass buttons, to the top, where a white collar circles his neck. The six of them stare into the camera.

I stare back, at one picture and then the other, for hours, at one face and then at the same face a year later. I see a year's difference; the children are older, bigger. But I can't say, other than that, how they've changed. Their round faces are no fatter, no leaner. I can't read the flat dark eyes, the closed mouths, the stiffly posed bodies. There is something, though, in the scrubbed look of three of the girls, in their mouths that are not quite smiling but held in a way that looks studied, like an assumed pleasantness put on with the clothes.

I know, of course, what the photos, in the government report on Education in Alaska, are meant to show. The poor waifs in their heathen hoods-up costumes have been remade in the course of one short year into well-groomed and civilized Christians.

Tumasock. Coogidlore. Esanetuck. Kokleluk. Anneebuck. Lablok. These are your names, Inupiat children, written beneath your photos in awkward English. Burroughs certainly did not meet any of you in 1899; you had already been removed to a place at least as foreign and inhospitable to you as the Arctic tundra was to him. Perhaps you were taken away as orphans, your families fallen to disease or depravity brought by whalers, miners, even by kindly American explorers who visited your shores. Perhaps you were removed from your homes in the well-intended but utterly mistaken belief that you would benefit from abandonment of who you were. You were told your people were ignorant and superstitious, your culture barbaric.

The founder of the Carlisle Indian School lived by this slogan: *kill the Indian, and save the man.*

I study your beautiful children's faces, light and shadow, unreadable eyes. Behind your Pennsylvania masks, I know there must be sorrow and shame. There must be rage.

TRANSCENDENCE

Before midnight, I get out of my rolling bunk to take a Dramamine, then make it through the leeside door to lean over the rail and vomit into the sea, that dainty yellow pill and all the green bile of an empty stomach. I let the cool air flood my face while I train my eyes on the gray-met horizon. Nothing but sea and sky that way, thousands of miles of unpacific Pacific.

When I at last turn back to the door, I see what I'd missed on my rush to the air. On the opposite side of the boat, to our north, the sun has just set behind a darkening island, and the sky all around the island is a pink the color of hair ribbons, of buoy balls, of the smooth surface inside conch shells. It's stunning in its brilliance, and in the way it outlines the island like hot coals and turns pastel and swirly as it rises through the sky. Even in— or perhaps because of—my indisposed state, I have an oddly comforting sense that the sunset is not at a distance, something to admire from afar, but *with* us. We are within the scene itself, under the glow, bathed directly by its light.

Burroughs was much taken with Alaska's sunsets; although he claimed that he'd often seen as much color in the sky, it was the way the colors accented and gave depth to the mountains and sea that was new to him. After a particularly flaming one that surrounded and played off of snowy peaks, he was moved to write, "The solid earth became spiritual and transcendent."

This was the oft-reported occasion when Burroughs called down from the bridge to Muir on the deck below. I hear the sharp play in his voice as he needles the friend who so often was the first to do the needling. "Muir," he calls, "you ought to have been out here fifteen minutes ago, instead of singing hymns in the cabin."

But Muir is quick, and he knows, too, where to find his religion. "Aye," he answers. "And you, Johnny, ought to have been up here years ago, instead of slumbering down there on the Hudson."

I try to see Burroughs's face in the fading light. Was he peeved? Surely the remark stung, reproof to the man who stayed home from the man who saw so much more of the world and chose the wildest, highest, and iciest parts for his own.

In our wheelhouse, as though one of us has called the rest, our crew quietly assembles before the sky. Some bring cameras, some late-night bowls of popcorn. We watch the show, all extravagance and all soothing tonic, until the light fades to palest pink and then blush and then a yellow streaked gray. We admire, but we don't much talk—only about camera film, about landmarks and weather and who will take what watch through the night.

The solid earth became spiritual and transcendent. Who would dare write such a line today? Who would risk such a grandiose Romantic effusiveness—without doubt, without irony, without fear of seeming silly? And yet, it is true that at such moments everything under the sun becomes transcendent, beyond all knowing. It is our loss that we do not speak the language anymore. I don't know that, without it, we quite know what it is we see.

Birds of Large Ideas

We *arrive in Chignik Bay* just as the morning sun crests a flank of abrupt gray mountain still piled at its foot with the run-outs of snow avalanches. We drop anchor among several seine boats—*Desert Storm, Miracle Girl, Desperado*—and offload the skiffs we've deckhauled from Homer for the fishery here, which opens in just a few hours. There is no harbor, just boats anchored in the bay, and the long dock that belongs to the cannery. The dock is crazy with activity—men and speeding forklifts, pallets stacked with seines, totes of ice, the cranes bending to boats rafted below.

The sun lays its golden stream across the sand beach and over the still water directly to our bow, as though we are the prize at the end of the line.

A half dozen large eagles hunker like tree stumps at the edge of the beach, close together, between us and the mountain and the sun. They are, I slowly realize, also captured in the beam of light, and not incidentally. Like domestic cats that will sit in sunny windows or under lamps, these birds have purposely positioned themselves for warmth. Other eagles and gulls flap slowly past, eyeing us, perhaps anticipating a good meal when the fishing begins and the processors pump salmon guts back into the bay.

I think of the *Elder*, passing along this coast, and of an occasion when several of the scientists stood at the rail and amused themselves by shooting gulls, seabirds, and eagles that flew alongside. They shot not because they disliked the birds, as fishermen who viewed eagles as competition would later shoot them for government bounties, but only out of foolishness, ignorance, the restlessness of men on a boat with too little to do.

Burroughs would not have been among those taking target practice on birds. Wherever he was on the ship, though, he must have heard the shots, must have known what was going on. Did he say anything? Did he scold loudly, demand or politely ask the participants to refrain, seethe silently, put his head under his pillow? Perhaps the group's host, Harriman, was among those with the rifles, and the host was not to be criticized.

Scientists. How did they feel, watching eagles collapse in flight and fold like wads of trash into the sea? What did they think—that eagles were like mosquitoes, they would always be there, no matter how many were wasted? In America in 1899, there were not yet many limits, not even to be recognized by scientists.

But Burroughs knew. He had long since stopped even collecting birds for specimens, and he had written of a despicable individual back in New York who killed deer only for sport and left the carcasses to rot: "There are traditions of persons having been smitten blind or senseless when about to commit some heinous offense, but the fact that this villain escaped without some such visitation throws discredit on all such stories." He left little doubt that, were he God, he would be more rigorous in his punishments.

From his home along the Hudson, he had once commonly counted a dozen or more of these same bald eagles in spring, but recently they had become rare enough for him to relish every one of his sightings. "The eagle is a bird of large ideas," he wrote. "I never look upon one without emotion; I follow him with my eye as long as I can." He understood the difference between knowing an eagle from its skin and eggs, and knowing one on the wing, finding inspiration in its "noble bearing" and in the idea of it in the sky, overseeing the tops of trees and mountains, the large view, the continent. *Dignity, repose, elevation*—these are the words he matched to eagle, that he held out as ideals, what man himself might desire.

I think he grumbled quietly, our mild-mannered fellow, when the men fired from the rail, when they dropped eagles into the sea.

CANNERY

A fearful smell, a big greasy cannery, and unutterably dirty, frowsy Chinamen. Men in the business are themselves canned.

MUIR JOURNAL

T*he good ship* Elder *stopped* at a cannery, not here but back in Prince William Sound. When a propeller blade broke on an iceberg, the captain beached the ship at Orca (today's Cordova) for repair, and the passengers marched off for a tour of the local industry.

This is an easy imagining: Burroughs so thoroughly impressed with the skull-capped "Chinamen" he met on the dock. He tells us, "It was positively fascinating to see the skill and swiftness with which some of these men worked. . . . There was a gleam of steel about the fish half a moment and the work was done." Those long knives, a few lightning-speed strokes. Fins, head, tail, innards—all sliced away in seconds before the neatly truncated packages were passed along for washing, then pitched to the conveyor belt.

Inside the Orca cannery, rotating knives plunged into the fish, cutting them into one-pound hunks that a mechanized rammer punched into the open mouths of cans. Workers soldered on lids, checked the closed cans for leaks, rolled trays of cans into retorts for pressure cooking. Labels were stuck, cases packed. Every second a new can of salmon came off the production line—all day, every day.

The sights and smells were too much for Burroughs, who lost his appetite entirely. Mountains of dead salmon were heaped in the sun, pecked at by birds, and all the waste lay under the dock in stinking piles that washed back and forth with the oily, bloody tides.

It is this waste Burroughs turned from, the spoiling fish and the guts and strong odors. I will him to see the larger waste, the abuse of an entire resource. I will him to outrage. He is blind to it, or merely reticent; perhaps, again, as a guest of Harriman's and here, of the cannery's management, he is reluctant to criticize.

Fortunately, someone else was taking notes. Grinnell studied the canning process, observed the mounds of fish and waste, asked questions. Later, in the report he writes for Harriman, he will indict the industry for its greed, for blockades at stream mouths that oftentimes robbed every spawner from a run, for the indiscriminate catch and disposal of unwanted species, for the dumping of fish when more were caught than the canneries could handle. He did not accept the boast that Alaska's salmon were inexhaustible, and he kept asking until he learned that, in fact, it took longer that summer of '99 to catch the same number of salmon than it took just a few years before and, yes, fishermen had to travel farther to find the runs. Salmon were, in fact, declining at a remarkably rapid rate.

The cannery men knew this, but they continued their wasteful and thoughtlessly selfish ways. Their motto, Grinnell said, seemed to be: *If I do not take all I can get somebody else will get something.*

That year, even the government was beginning to express alarm. The territorial governor would write Washington to ask for limits. He would pass along petitions from Native people, complaining that they were being kept from their ancestral streams, that their fish were taken away in shiploads, that they would starve. One of these polite petitions reads, in part, "We tried to remonstrate with them [the white men fishing for canneries], and they threatened to smash our skulls."

I want to shake Burroughs, official trip historian. *Look* at what's in front of you! Tell the world about it! Condemn the waste and injustice! *Enough* about the admirable skill of the Chinese fish cutters! I look for him and he is not even on the dock anymore. Our man has left the disagreeable cannery and polluted waterfront behind and is off to his sweet-smelling woods, following the song of a hermit thrush.

ALL THIS

In *Chignik*, the cannery is short of tenders for the first opener, so we sign up for a one-day charter. From behind a sand spit, all day in glorious T-shirt weather, we watch a handful of seiners make their sets, circling near shore or holding hooks before closing and pursing. The boats are freshly painted in brilliant whites and blues; the bumpered skiffs glint; fishermen in orange rainpants work the back decks, restacking corks and web that churn through the power blocks above their heads. The pace is steady and leisurely, practiced; there aren't a great many salmon yet, no schools visibly darkening the water, no frothiness within the pursing seines, not even any jumpers. Sound carries clearly across the water: the rhythmic plunging, the steady grind of hydraulics, the clink of purse rings, voices. Ken watches with binoculars as one boat finishes drying up its net and rolls a load of flashing-silver fish aboard. The water in the bay is so smooth it might be oiled, and the color is Mediterranean.

Around the bay, waterfalls pour down steep rocky cliffs. The hills above are still greening, still laced with snow. Eel grass swishes like streamers in the shallows, and all along the sand spit piles of wave-beaten cobblestones and bleached logs lie like polished bones.

This is the romance of fishing: the perfect day, the calmest sea, a place on earth more lovely than most people can even imagine. Honest work, muscled pleasure. All this, and there's money to be made, too?

Forget storms, web wrapped in the wheel, groundings, a screaming skipper, slacking crew, failed sets, broken hydraulics, no fish, boat payments, jellyfish dripping into your eyes, nets full of logs and baby crab, days of rain and dry heaves, your own numbing exhaustion.

Our first delivery comes after dark, under a full moon that has risen over a high mountain shoulder to bathe the circle of mountains, all the rocky beaches, the entire bay, in a solid creamy light.

The *Jerilyn Dee* is a family boat—mother, father, daughter, a couple of young boys hanging in the cabin door—local people, like most who fish here at Chignik. They tie up alongside, and our crew gets to work under the sodium lights, moving the pump hose to the boat's hold, readying the scale. Ken speaks to the pleasant woman—talks prices, fish care, how's it going? She pulls off gloves, finds her permit in a shirt pocket to hand him. Through the seiner's galley window, I watch the younger woman juggling pans and plates. No one is in a hurry here, not now, late at night under a full moon, on the first day of the new fishing season. The man takes off his hat, rubs the back of his neck with a big, dark, muscled hand. He says something admiring about our tender's setup, about our refrigerated seawater. Ken asks if they need ice, supplies, anything at all. A smell of soup and hot grease seeps from the smaller boat's galley, across their deck and ours. This is not just a business transaction, here, the sale of fish; it is the tradition of visiting, that timeless village way of being.

Had the Harriman expedition stopped here on its passage out the peninsula, Burroughs would have met cannery workers like those he saw in Orca, imported from San Francisco's Chinatown, the Philippines, Hawaii. The fishermen, sailing and rowing their beamy boats, would have been Scandinavian, Italian, Greek. The Native people of this place—Pacific Eskimos with their distinct and prosperous 5,000-year-old culture—may not

even have been visible, squeezed into the margins of their for-
mer lives, robbed of traditional foods, forced to the woods to
trap ever-depleting animals for the sale of their furs.

But Burroughs would also have seen people standing idly
about the half dozen canneries in the bay and lagoon, the workers' knives at rest. In 1899 the Chignik salmon fishery was a fiz-
zle, the runs wiped out. Chignik had been the source of a major
portion of Alaska's canned salmon production, but the cannery
bosses and fishermen had, by the usual practice of setting nets
across river mouths, exhausted the local resource in just a few
years.

Today, Chignik's salmon fishermen are among the most prosperous in the state, year after year, in a carefully managed fishery. The heritage of these people goes deep—a far, knowing
intimacy into the lives of salmon and with the place their language knows as "big wind." But they have other heritages, too. I
look down the list of fishermen the cannery gave us, and it reads
like a Swedish-Norwegian phonebook, with a few Russian and
other names mixed in. Those Scandinavians who came with the
cod fishery, and as settlers, and to fish salmon, brought their own
passions for the sea, their desires for a better life. Yes, it was
colonialism. Yes, it amounted to cultural genocide. But, also, yes,
the assimilation worked both ways. Those who came from afar
became a part of the place, adapted and adopted and married in,
became family.

Drive

There is one word of advice and caution to be given those intending to visit Alaska for pleasure, for sight-seeing. If you are old, go by all means; but if you are young, wait. The scenery of Alaska is much grander than anything else of the kind in the world, and it is not well to dull one's capacity for enjoyment by seeing the finest first.

HENRY GANNETT,
"General Geography," *Harriman Alaska Series*

F*rom Chignik, early morning*, I drive the boat west while the others sleep. This is grand and gorgeous country, writ large. Castle Cape rises in sun-warmed red rock like something that belongs in the American Southwest—sheer sculpted towers lifting from the sea like, indeed, castle turrets. Its sedimentary layers circle and fold in wider and narrower, lighter and darker bands. I reflect on the fact that it was only the year before the *Elder* passed that this astounding landmark was described for the record for the first time—and not until 1926 that it was officially named. It had a name before that, of course (Tuliumnit Point, the meaning of which I don't know—certainly nothing as Old World as *castle*), but I'm impressed at how "new" this coast is to American geography.

Our course is plotted on an electronic map before me. I need only steer to the compass reading that blinks at me in green light. And not even steer, really—not by turning a wheel, but just by directing the rudder with a little dial. The whole exercise is

like connecting dots, child's play. It requires no skill beyond staying awake and watching for floating obstructions—unlikely in
this treeless land. I check the fathometer for depths, fret mildly
over possible hidden rocks, turn wide around the various points
and capes that jut from the peninsula shoreline.

I can scarcely imagine the abilities—and desires—of the old
steamer captains who ran this shore in fog, storm, and winter
dark, with only compasses, lead lines, and their own sharp senses
to guide them. When there were no landmarks to make out, they
ran full speed ahead, measuring distance against time; any slow-
ing and they'd be lost. They blew their whistles and steered by
the echoes bouncing off the mountainsides, by the sounds of
waves breaking on shore. They "smelled" their way along—
sniffing plant resins, mud, the decay at fish streams, feeling the
geographically induced wind and temperature shifts on their
skin. Their charts, when there was anything marked at all, were
sprinkled with notations of P.D. and E.D.—"position doubtful"
and "existence doubtful."

I steer another degree south, see on the screen our passage
defined by another tiny dash past Seal Cape, toward Mitrofania
Island. I am in awe of the men who truly navigated this coast and
the technology that allows such a fraud as I to sit in a captain's
chair.

The entire morning, plowing through blue seas dotted with
pairs of parrot-headed puffins, I see no other boat, no plane in
the sky, no person or sign of person on land or anywhere at all.
We are the small mote of dust in a galaxy, the drop in the sea, the
inconsequential flicker through a timeless landscape.

Memory

The endlessness of this country, all that distance. And the seabirds that stroke past, soot-colored with long, crescent-shaped wings—jaegers, petrels, shearwaters, I don't even know. It's enough to see another white peak rising in the light, and the auklet that turns its yellow eye on us.

I remember something I have not thought of in years and years. When I was a very small child, before I was old enough for school, I used to imagine that a movie camera spun behind my eyes, and that everything I saw I was recording and then projecting onto a screen so that all people everywhere might see it, too. Everything then was so spectacularly new and intriguing to me, I was convinced that everyone else in the world would be equally fascinated to share my view. I took particular care to hold my eyes wide open, to seek out the most interesting things and to look carefully at them, to keep that camera rolling. Clouds in the sky, the way ants walked in a line across the sidewalk and carried grass seeds as big as themselves, patterns on wallpaper and silver forks—all of life's small details and discoveries were wondrous to me, worthy of my most careful documentation.

In my child's imagination, I could turn the camera behind my eyes on and off. I turned it off out of modesty when I went to the bathroom, but the rest of the time, for hours and days on end, I was conscious of directing an ongoing pageant. I remember running that camera in my bedroom, staring through the whirring blades of a fan set in the window at tree branches beyond it, the green of waving leaves, the smudgy artistic effect of the moving blades.

I am there again, in that place of all beauty. The camera behind my eyes is rolling, zooming in and out, capturing every angle, using the light to its best advantage. What I see all around me—from ocean to hills to untracked sky—is as new as the start of life on earth, as fresh as a baby's first open-eyed look at its mother. All the people in all the world should see what I am seeing. If they saw it, they would love it, they would know that life is filled with both many small blisses and stark raving extravaganzas, well worth taking whole-hearted part in. I am as self-assertively sure of this now as I ever was.

ALBATROSS

Never before had I seen flying so easy and spontaneous,—not an action, not a thought, not an effort, but a dream.

BURROUGHS,
"Narrative of the Expedition"

ithout my bird book or a Dall or Muir to ask, I am guessing, but I think the large, dark, heavy-billed bird following us, casting its eye upon me as it swings, with barely a tilt of its long, slender wings, past the pilothouse, is that most mythic of birds, an albatross.

On the Gulf of Alaska, crossing from Yakutat Bay to Prince William Sound, Burroughs saw his first albatross, that flight that was nothing but dream. "It seemed like the spirit of the deep taking visible form and seeking to weave some spell upon us or lure us away to destruction."

Burroughs would have known Samuel Coleridge's "The Rime of the Ancient Mariner," would have thought on seeing the albatross of the ancient mariner's sin in killing the harmless bird, and his punishment—not just of his shipmates hanging the dead bird around his neck but of the fates becalming his ship. Coleridge published his most famous poem in 1798, one hundred years before Burroughs was to see his own albatross. And here I am, another hundred years later, with the same bird following in its same venerable silence, and both writers speaking from the back of my head.

I wonder how many people today have never seen an albatross and yet might recognize one if they did, based on a two-hundred-year-old poem? Certainly, millions of people who have never read the poem or heard of Coleridge nonetheless know what it means to have an albatross hung around one's neck, so deeply has that literary reference permeated our general culture. So too the famous lines about water, water everywhere and not a drop to drink and the wisdom of caring for all things great and small—we all know them, if not their origin.

I wonder something else, though, about Burroughs in his poetic recall. Would he have known that Coleridge, when he wrote his great poem, had never been to sea? The poet had grown up six miles from the coast and had seen the ocean from the shore; he also, apparently, was an avid reader of sea voyages. That combination was all he needed to invent a detailed voyage to Antarctica and back through the storm latitudes and trades into the equatorial doldrums. The "esemplastic power" of the imagination, Coleridge himself called it: the ability of an artist to mold, shape, or fashion odd bits of whatever into something altogether new and fabulous. Such imagining has little to do with the power of observing—Burroughs's great strength—and everything to do with reaching down into the depths of one's mind, past all reason. We can know things beyond any experience or explanation.

These are facts: Coleridge's albatross was the southern species known as the Wandering Albatross, the largest flying bird alive today, with a wingspan of more than eleven feet. Of the three species found off of Alaska, one is endangered, one is considered rare, and only the black-footed albatross—the bird I decide I must be seeing, the one Burroughs saw—is commonly spotted from vessels at sea. The North Pacific species were exploited on a very large scale during the last half of the nineteenth century (up to H.A.E. time) by the Japanese for their feathers.

The main diet of most albatrosses is large squid seized at the water surface, though they also scavenge after fishing boats.

A dead bird around an old sailor's neck, on a voyage that never was: how strange to find such a thing most recognizable, most commonly acknowledged, perhaps truest of all.

GREEN

The volcanic *Shumagins* part and interlock around the sea like odd-shaped pieces of a puzzle. They rise at their edges with vertical abruptness, then slope more gently uphill and along clean-cut ridges. Their topography is everywhere as plain and uncluttered as a relief map made from clay. What you see is what you get, these fifteen main islands mixed with the many more smaller ones, like stepping stones, among them.

And what I see, in the lowering light, is more fiery Irish green, spilling like liquid down the slopes, color seeming to drip from rock, to catch and concentrate on every level surface. It is that same green about which Burroughs waxed all the way from Kodiak, this way and beyond. He adored this country for its—as he saw it—pastoral splendor, all the "smooth rounded hills as green and tender to the eye as well kept lawns," all the "sweep of green skirts," "green carpet," "vast meadows," for "suggesting endless possibilities of flocks and herds and rural homes." "Green as a lawn," he says again, and again—five times I find Burroughs comparing this treeless green country to tended lawns.

I see the same green splendor, the same openness that Burroughs saw, and I adore it, too—for entirely different associations and near-opposite reasons. I look upon these achingly green islands and see not lawns and farms, nothing tame or domesticated, but wildness. What I see is seamlessly green and tirelessly unrolling, untracked by man or woman or domestic beast, not tended, not mown, not made "useful."

Burroughs spent his life trying to re-create a rural past, both on his small farm and in his books. According to his biographers, he felt a tremendous, painful nostalgia for the rural life he had known as a boy, and for what he thought was his country's enviable and irretrievable agrarian history. But his was an overboard nostalgia, romanticized to the point of being depressive. When Burroughs's son was born, he wrote in sadness, "I look upon this baby of mine and think how late he has come into this world—how much he has missed; what a faded and delapidated [*sic*] inheritance he has come into possession of."

Which is—something that comes to me with a shock of recognition—so similar to what I've always felt. Not a longing for a rural or farm past, but for an even older and rarer time. Growing up, I always felt that I had missed out by being born much too late to have known real wilderness. I dreamed of canoeing on lakes before they were surrounded by summer homes and zoomed over by speedboats, of standing on mountaintops and seeing nothing but trees and more mountains, of walking through hidden valleys to discover canyons, cliffs, hot springs that I had never known could exist. Wild pigeons, buffalo herds, big cats, tall waving prairie grasses—I wanted it all. I wanted to be Lewis *and* Clark *and* Sacajawea.

Burroughs and I are so different in this—or maybe not. Maybe only in the particulars. He, in his age, had his romanticized lost world; I, in my age, have mine.

But I have to ask: what does this longing say about *me*? About my view, pessimistic or otherwise, of the present and future? What is it I *really* want, or lack? Am I just a hopeless romantic, as unable as Burroughs to recognize the truths of the real, present world? Or might my yearning lead somewhere? Because, in fact, I don't want to return to an earlier time; I especially don't want to join the Harriman women in their corseted restrictions.

Right now, I only know what makes my heart glad—not even my heart, which, after all, is only a muscle, but that undefinable space within me, where the human spirit lives.

I can scarcely take my eyes from the green. When I do, the color, like concentrate, is still with me, and every other color a part of it, a pigment in the mix. Before us the flat blue water breaks at our bow, like satin being sheared with a blade, and pairs of puffins with glaring yellow bills paddle to one side or the other, or dive from our path.

FOXES REDUX

We *steam through* Unga Strait and toward the Pavlof Islands, in procession. Other boat traffic has appeared as if from nowhere, all lined out on the same westering course. We are behind a power scow, ahead of an enormous yellow barge with a towering load and crane. When the distant barge angles away, we see that its top is stacked with Bristol Bay gillnetters, four rows of five, their spars and rigging spearing the sky. The gillnetters must be regulation length—thirty-two feet—and nearly as wide, but they look like toys to be floated in a tub. Several seine boats come along behind us or fall in in front or keep company off to one side. We are close to Sand Point, the largest community in the region, and the home boats are headed out to their fisheries.

Ahead, on the island called Ukolnoi, stands a pitched tent too small for me to see, though the people working around it will later tell me they watched the same yellow barge with the load of gillnetters pass that day. They are people I know from home, Ed and Nina, and they are on a mission. They are killing foxes, trying to eliminate every last bird-eating one.

Ed and Nina have a dream, as passionate as fox-farmer Washburn's was for turning Alaska's "unused" and otherwise "useless" islands to good purpose. They dream that one by one Alaska's islands will be rid of introduced foxes, and that the pre-fox fecundity of those places will be restored. In their dream, the islands thunder with honks and chirrups and beating wings. Ground and crevice-nesting birds crowd every hillock, depression, and rock pile. The cliffs are jammed, ledge to narrow ledge, with webbed feet and pushy wings. Puffins pour in and out of their burrows. At night, the ghostly storm petrels flutter about like clouds of mutant moths. Ancient murrelets, red-legged kitti-wakes, whiskered and rhinoceros auklets, northern fulmars, pelagic cormorants, pigeon guillemots—these all are present. Geese and common eiders. Song sparrows and Lapland longspurs, and winter wrens, too. Peregrine falcons and snowy owls. Rock ptarmigans. Dozens of bird species. Hundreds of thousands of individuals. Millions. When they fly they blacken the sky as passenger pigeons once did over eastern forests. And still there is more. Grasses and ferns and cascades of bright flow-ers, luxuriant in fertile soil, overflow the islands. The cliffs drip with chalky excrement, enriching the surrounding waters, which in turn teem with plankton, small fish, bigger fish, sea lions, and whales. The plankton at night is luminescent, swirling like glit-ter in the currents, in the passage of more diving birds. Richness, diversity, mind-numbing numbers—here is a kind of evolution-ary perfection.

In this dream there are no introduced predators, not a single land mammal except the occasional river otter. There are only the raptors that swoop from above: egg-stealing gulls, the flashy parasitic jaegers.

This dream looks a lot like other places Ed and Nina have been—real places, remote islands that have so far been spared invasions by foxes, shipwrecked rats, cattle, and other aliens. One small island they have visited, far out the Aleutian chain, hosts more than three-and-a-half million breeding birds of at least thirty-two species. One percent of the Aleutian islands land mass—seventy percent of its nesting seabirds.

Think of that. Think of the possibilities, of what once was.

To be sure, foxes lived on some Alaska islands before the in-tervention of people—foxes that had crossed from the mainland on winter ice or perhaps been left by ice-age juggling. On Vitus Bering's 1741 voyage of discovery, in which he claimed Alaska for Russia, he named the Fox Islands in the Aleutians for what he found there. Russians introduced foxes from place to place, and by 1811 Aleuts were already complaining that they could not find enough birds on their home islands to make their bird-skin clothing, that they were having to make dangerous journeys to farther, foxless islands.

Little could the members of the Harriman expedition have known that the summer they passed through Alaska was the beginning of fox farming's golden age. Like a pyramid scheme or a plague, pairs of foxes were sent forth with the territory's newest entrepreneurs, destined to be fruitful and multiply. By the 1920s fur production was Alaska's third largest industry, and fox farms had taken over *four hundred fifty* of Alaska's previously fox-free islands. On some treeless islands, virtually every bird, except those that nested on unreachable cliffsides, was eliminated.

The fox farming industry collapsed with the Great Depression and never recovered. History tells us this, as history has left the abandoned farms with their fallen cabins and tilted fence posts, the fox cages swinging on their hinges. On many islands history has left foxes, too, in whatever numbers food supplies and inhospitable environments can support. What history does not tell us—because there is no record, no pre-fox inventory—is what existed before the foxes, what is now gone, perhaps forever. We only know a larger history—that, worldwide, 70 percent of the bird species known to have become extinct were native to islands and were lost because of the introduction of exotic species.

In Alaska, the Aleutian Canada goose was thought for a time to be extinct, wiped out by foxes. Only in 1962 and again twenty years later were remnant flocks discovered, and only a program of fox eradication and habitat restoration has allowed populations to rebuild.

And so we pass on down the coast, by the camp where Ed and Nina volunteer some summer weeks to an eradication effort that began in 1949 and continues, island by slow island. Among those islands that today make up the Alaska Maritime National Refuge, foxes have so far been removed from twenty-nine and remain on at least thirty more. There is no efficiency in this effort. Poison is efficient, but poison is strictly against the law. This battle is waged with traps and rifles and cunning. When their time is up and Ed and Nina leave Ukolnoi Island, they will leave knowing that there was one wily fox—they hope *only* one—that still got away.

NEST

O*n its passage* through the Shumagins, the *Elder* stopped at Sand Point, now a prosperous fishing town of a thousand residents, then described by Burroughs as no community at all but a harbor with "alder-clad shores." A small party was left off to study the volcanic formation of the land and collect marine specimens. Sand Point had been the meeting place of the fur seal poaching fleet, later a supply point for ships sailing between San Francisco and the codfishing grounds off Russia, a trading post, a buying station for salmon, and a fox-farming center. But in 1899 a single caretaker lived among the crumbling houses and the hotel where the scientists would camp.

During its stop the expeditioners tramped all around the old settlement, the beach, the hills, collecting as they went. More birds, butterflies, pink primroses, hunks of rock, some of the several hundred new crustaceans gathered by expedition members and eventually inventoried.

Back on the ship, sometime after it escaped a couple of drunken miners chasing behind in a rowboat, came Burroughs's most uncomfortable moment of the cruise. He, understandably, says nothing of this in his account, but the story nevertheless became part of the H.A.E. legend. It goes like this:

Chief scientist Merriam is sitting on deck, admiring the mountains, when Burroughs sidles up. Burroughs has in hand something he collected at Sand Point—a fox sparrow's nest with four eggs. He doesn't want the others to know, not after he's complained about their collecting, but would Merriam be willing to make a private trade—the nest and eggs for the skin of a golden-crowned sparrow he'd taken earlier?

Burroughs apparently had been admiring, coveting, the deliciously bright-capped skin for days, perhaps weeks. Among the little brown birds in the sparrow family, the golden-crown is certainly among the loveliest in looks, with its flame of color capped between dark bands that pass like heavy eyebrows on either side of the bird's head. Burroughs would never have seen such a bird before this expedition; golden-crowns live only in the far west, wintering along the coasts of California, Oregon, and Washington and breeding in British Columbia and Alaska.

The poor man. I think he was seduced by the bird's jeweled beauty. He might have gone for any number of dramatic Alaskan birds, but no—he wanted only this, this one small, elegant specimen. He had found the object of his aesthetic desire.

I don't know if, cruelly, Merriam refused to make the deal. I do know he laughed at the old naturalist and went on to announce the proposal to the other scientists, to let them all know that their critic was not above collecting after all. And that after that the whole company teased Burroughs unmercifully, memorably enough that Kearney, one of the young botanists, would highlight the event in a reminiscing letter a full half-century later.

The Chart, Assisted
in Its Reading by the
Dictionary of Alaska Place Names

W*e are past Kupreanof Point*, Karpa Island, the Shumagins. The new coastline folds, sheers, breaks into dramatic pieces. I study the chart for perspective, to pick out points and bays, islands and mainland.

The early Russians left their naming everywhere: Wosnesenski Island, Chicagof Bay, Poperechnoi Island, Dolgoi Cape, Olga Rock, Belkofski Bay, Cape Tolstoi.

Wosnesenski, the name, has a homey familiarity. Back where we began, in Kachemak Bay, Wosnesenski Glacier and Wosnesenski River spill down out of the mountains. The man, Ilia G., explored there and here for the St. Petersburg Academy of Sciences a century and a half ago. Long before that, this island was well known to the Aleut people, who called it by a word that referred to its crested peak.

Chicagof Bay reminds me of Chichagof Island in southeast Alaska. Sure enough, they are both named for Russian Admiral Vasili Yakov Chichagov, who explored Alaska even before Wosnesenski, way back in 1765–66. The H.A.E. reported Chicagof as the "local" name here and had it officially adopted.

Poperechnoi, Dolgoi, and Belkofski are all descriptive names. The first two are Russian for "sideways" and "long"; the last alludes to an abandoned village named for its squirrels.

Olga, the wife or sweetheart who had her very own rock named after her, is lost to history. But again, a link to the H.A.E. Expeditioner Dall reported this name on his 1882 trip to the area. (Dall, of the porpoises and the sheep, had done his own female-personage naming—an entire large island in southeast Alaska for his wife, Annette.)

The cape named Tolstoi, I regret to discover, was not named for the great novelist and social critic by a literary-minded ship captain. *Tolstoi* is Russian for "broad."

Neither is Apollo Mountain named for the god of music, poetry, and manly beauty. It is named instead for the Apollo Mine, which for a brief time scratched unprettily at its side.

The grandeur of this landscape is, nevertheless, recorded all across this end of the chart. Dolgoi and Tolstoi, and, in English, Pinnacle Point, Monolith Point, Cathedral Peak, Castle Rock, Elephant Rock, Arch Point, Long Beach, Volcano Bay.

I find, too, Slime Bank, Mean Rock, Clubbing Rocks.

I study the marked navigational hazards—the symbols and the words themselves in fine print. They fill entire bays and false passages, a geography of disaster. Rocks, reefs, breakers. Lava and ashes. Tide rips and strong currents. The simple word, *foul*.

At the far end of the day, we reach the twin volcanoes, Pavlof and Pavlof's Sister, with the sun setting in pale orange behind them and a flat, mushroom-shaped cloud hovering over Pavlof's peak. Both white cones are at rest, though they will vent steam and erupt ash again—as Pavlof does every several years, as it was doing when gazed on by the H.A.E. In the same slow twilight, we slide past the Aghileen Pinnacles, stark rock spires towering gray and vertical at the ridgeline of snow-covered mountains, like the ruins of an ancient, fortified city. The pinnacles are one of the most spectacular sights on the peninsula, or anywhere, and we're lucky to find them both clear of weather and backlit, as impressive as they must ever be. Aghileen—a gorgeous name, and another connection to Dall; in 1880 he had reported this name as an Eskimo one, but its translation, if he knew it then or two decades later, has not survived.

Ken says, "Every time I go by here I wish I could hook Anchorage to the stern and drag it past with me." He means, I know, that he thinks if the people in the city could only see this wondrous country, they would better appreciate where they lived and care about it more. They would have to change their lives.

A few more points to pass, islands to dodge—we are coming to the end of our sail, four days from Homer, some six hundred miles, here to the far squiggled end of the Alaska Peninsula. Where the peninsula ends and the Aleutian Islands begin, Isanotski—the Russian corruption of the Aleut *Isanax*, meaning "the pass"—is the first place that water flows between the North Pacific and the Bering Sea. Water and the things that travel in and on it—salmon, whales, and boats.

Captive

The Harriman expedition went farther, past the big island of Unimak and several small islands to the island of Unalaska, where they docked for water and coal at the new town of Dutch Harbor. The North American Commercial Company, busily decimating fur seals, kept offices there. Steamships full of delusional gold miners stopped there before heading for the beaches of Nome.

There the *Elder* would pass from its coastal tracings and strike out northward across the open Bering Sea, all the way to Siberia.

Unalaska looked pretty good to Burroughs. It was green, it bristled with ever-newer and more luscious wildflowers, and, best of all, it stood on solid ground. He had savored his strolls around town almost as much as those in Kodiak, and he had even found a kind woman who would rent him a room and feed him fresh eggs until his ship returned.

There he is, satchel in hand, quietly descending the gangplank.

And here come Muir and his cabinmate Charles Keeler, appearing out of nowhere. Muir asks, with shock in his voice, where Burroughs could possibly be going. He certainly cannot be leaving the ship. The best is yet to come! Muir ticks off on his fingers the joys to be found in the Bering Sea. Fur seals breeding in millions on the Pribolofs! The Eskimos at Port Clarence! The midnight sun! All of this must be captured by the expedition's historian.

Burroughs says something about the tempestuousness of the sea.

Muir promises: the Bering Sea is a millpond. He takes Burroughs's arm, Keeler his satchel, and the two escort their friend back on board.

Muir account: "We kept John Burroughs on the ship."

Keeler account: "Mr. Burroughs did not want to go into Bering Sea, but he could not stand Mr. Muir's scorn. He weakened a bit and was lost."

Burroughs's account: "If only I could have a few days of that kind of intimacy with the new nature . . . [but the ship left for the Bering Sea] and I was aboard her, with wistful and reverted eyes."

ROMANCE

*Later, when we had been aground on a rocky reef off the Pribolofs
and had tossed about on the fog-hung, tempestuous waters of
Bering Sea, Mr. Burroughs lay in his berth and groaned . . .*

CHARLES KEELER,
unpublished manuscript

I *of the weak sea stomach* have no trouble imagining this:
Poor John Burroughs lies ill in his bunk, his every
muscle exhausted from bracing himself against the ship's roll.
His stomach aches from the violence of his heaves; his throat is
raw. Up and down, up and down, at the mercy of those monstrous, inescapable waves. Spittle has caught in his beard, and his
eyes, hooded and red, weep with his own pity. Now he buries his
head in his pillow. If he were on deck, he thinks, it would not
take any courage at all to throw himself over the side—death by
drowning in frigid water would feel better than these miserable
unending days crossing the vile Bering Sea.

All because of that madman John Muir. Muir, who was never
seasick one day of his life and who climbed trees just to rock in
their tops in tempest-storms. Let Muir write the trip's narrative
then! How much can anyone say about water, water everywhere,
all of it dark and empty and going up and down in swells you
could lose a whole mountain behind?

A knock at the door. Keeler takes his seat beside the bunk, offers a drink of water, opens a small leather book.

This is his task, commanded by Muir and good conscience: to look in on Burroughs and see to his comfort. And comfort, according to Muir's orders, involves the words of that great Romantic poet, William Wordsworth, who knew all of nature to be of great beauty and sublimity. This I know: Muir had carried Wordsworth's words—and Yeats's and Blake's—into the American West, where they had tinted his every view of every river, lake, valley, and mountain. Muir found not want, but rhapsodic inspiration, in wild places. It was this vision, as profoundly felt as any religious belief, that led him past worship to politics. He was not content to merely praise; soon he would be getting after a new president, Teddy Roosevelt, to protect his beloved Yosemite Valley, and he would devote his last years to trying to save Hetch Hetchy Valley from the drowning waters of thirsty men.

Burroughs, too, has labored under the influence of the Romantic poets all his life. When Keeler reads, Burroughs lets the familiar words, the rhythms, wash over him like a counterwave, a gentling current. He concentrates on the falling off of Keeler's voice at each rhyming break, on images of stars and trees.

Beside him, on the rimmed table, his own reading material slides back and forth. What might Burroughs have brought aboard for reading at calmer moments?

We know the major influences of his younger days. Emerson's essays, to begin with. Early on he took to heart two of Emerson's rules for attaining true culture: sit alone and keep a journal. His first bit of juvenile writing, submitted for publication, had been mistaken for work by the great man himself.

Thoreau he read and admired, though he found him "grim, uncompromising, almost heartless" and thought his own powers of observation sharper. He also tired of comparisons that were always made between himself and Thoreau. (Among them Henry James: Burroughs was "a sort of reduced, but also more humorous, more available, and more sociable Thoreau.")

Back in his Washington, D.C., bank vault–guarding days, he had followed after the older Walt Whitman like a pup. His first published book, the hero-worshiping *Notes on Walt Whitman*, was also the first book written about Whitman. "Whitman does not to me suggest the wild and unkempt as he seems to do to many; he suggests the cosmic and the elemental," he wrote, putting his civilizing spin on *Leaves of Grass*.

Then there was the blasphemous Darwin, whose *The Descent of Man* had reinforced his beliefs about man being a part of, not separate from, nature.

Among his contemporaries Burroughs preferred those with a similar "rambling" style of nature appreciation—Bradford Torrey, Dallas Lore Sharp, Rowland Evans Robinson. He was not a fan of John Charles Van Dyke, who argued for wilderness preservation, or of Winthrop Packard, who proposed laws to protect nature from being overrun by nature lovers. Van Dyke and Packard, with John Muir, had been strongly influenced by George Perkins Marsh's 1864 book *Man and Nature*, regarded as the granddaddy of American conservation literature. Burroughs is not known to have read it and was not, in any case, interested in its arguments about the values of wilderness.

But for now, Burroughs is too ill to read anything on his bedside table. He suffers being read aloud to, perhaps this day Wordsworth's "The daffodils":

> *I wandered lonely as a cloud*
> *That floats on high o'er vales and hills,*
> *When all at once I saw a crowd,*
> *A host, of golden daffodils;*
> *Beside the lake, beneath the trees,*
> *Fluttering and dancing in the breeze.*

On and on—all those bright daffodils engaged in sprightly dance, continuous as the twinkling stars, outdoing the sparkly waves, filling the poet's heart with pleasure.

Wordsworth was, unlike his friend Coleridge, a sailor—but one who confined his sailing to lovely English lakes. He would not have gone larking about on the punishing Bering Sea.

I think the irony must make poor Burroughs groan.

PART III

FISHING GROUNDS

H *ere and now:* the hard rain falling, and the hills dim and gray with a yellowed, washed-away cast. The seiners and gillnetters anchored around us, riding out the weather within Ikatan Bay. A kittiwake chopping past. The anticipation.

Today is the first opening in Area M, and there is not a soul for hundreds of miles who doesn't know this, who isn't involved in some intimate way with the fishery that sustains this region and its families. Beyond salmon, beyond weather, beyond testing new gear and crew, there is nothing that matters. Four hundred boats fish here, and for everyone on them, and on the tenders and processors, there is no other world.

From the pilothouse I watch through blurred windows as our crew tears apart cardboard boxes and loads resale goods into our freezer and storage. Steaks, ice cream, bags of bread and jars of peanut butter—these, too, are part of our service to the fleet. Beside and above and behind me the multitude of radios booms and cackles. I am to listen for anyone calling us, for weather, for news of catches, prices, species ratios. The two VHFs are set to different channels; the single sideband can leap voices over mountains; the mysterious black box, on its private frequency, communicates solely with our company and its other tenders.

Fish buyers are calling through the channels, advertising prices, bidding each other up. $1.30. $1.35. $1.40. Cash. Bonus later. Everyone talks reds, no one wants to mention the "other" fish, chum salmon. The real news, we know from the company, is that too many chums are being caught. The chums are meant to get through for fisheries farther up the coast. Too many intercepted here, and the whole fishery will shut down.

On the radios, fishermen growl among themselves in protective code. Their locations are vague, their catches shared in senseless syllables, in phrases such as *coconut pie* and *betty to veronica*, in Russian. We, too, will be reporting numbers of boats and pounds in a scrambled code provided by the company office. Zulu boats. Hotel alpha foxtrot thousands of pounds.

From the sheltered bay, we are sent back out to the fishing grounds. Ken navigates with binoculars held to his eyes; there are boats everywhere, and it can be hard to see and avoid their nets in such "tempestuous" (to borrow a Burroughs word) seas. Few boats, though, are actually fishing at the moment. They are running from one place to another, or they are waiting—for calmer seas or more fish, preferably both.

In East Anchor Cove, the anchored boats all lie with their bows into the wind, like birds of a flock. We easily locate our friends on the *Lucky Dove*. Buck and Shelly and their two crew members are in from tending their nets off the south side of the peninsula, where they had all gotten sick in the rocking and rolling.

A few pints of Häagen-Dazs from our freezer, and they are revived.

Later, when the weather calms, we motor out to their sites to collect fish. It is an exposed, rocky shoreline they fish along, with the sea crashing in and long, gray, lost-in-the-clouds hills behind. The *Lucky Dove* serves as shelter for eating and sleeping, but their fishing—set gillnetting—takes place from small open skiffs that can work close to shore, the fixed nets drawn across their gunnels. We watch from our anchorage as the nearer skiff tosses in the waves, the orange-hooded figures bent over the net, pulling and picking, snapping salmon from the web: a very small tableau, pinpricks of color, against a dramatically large and somber landscape.

This is a romantic vision, yes, that we cling to, men and women working the sea in this time-honored fashion. Not many do it like this anymore, one fish at a time. It is not the most efficient way, certainly, nor the easiest or safest. But the salmon fisherman knows an art, and he knows about weather and whales and plankton blooms—all those connecting things that make up the salmon's world, and his own.

What is it that Burroughs said? "Knowledge is only half the task. The other half is love." He was talking about writing natural history, about knowing facts and not stopping with them, but he might as well have been talking about fishing.

North and south of here—in the Bering Sea and in the Gulf of Alaska, modern, industrialized fishing ships are also at work. Giant factory trawlers, dragging enormous nets that can catch 400 tons of fish in a single tow, fish not for salmon but for groundfish like pollock and cod. They tear up the sea bottom, they waste enormous amounts of fish of the wrong sex, size, or species (including salmon) that they catch accidentally, and when they have fished out one area they move on to another. Little understanding beyond mere technical knowledge is involved, and, I think, no love at all.

My friends Shelly and Buck have noticed that when the pollock season opens in the Bering Sea, unusual numbers of sea lions pass through the strait beside their home, heading south, away from the fishery. The animals, now listed as an endangered species, appear to be either fleeing the disruption caused by the huge fleet of factory trawlers sweeping the area, or searching for food to replace the pollock removed so precipitously by the fleet. It is a fact that sea lions and other sea-feeding animals have fared poorly in recent years in areas of heavy trawling. It is another fact that today's scientists—specialists all—are reluctant to draw any conclusions. Each expert studies his own little piece of the ever more complicated ocean environment, and no one ever knows enough.

This is my sober thought: we have come so far from the time when a William Dall could excel not only in a broad range of the sciences but also in history, geography, anthropology, writing, and understanding. Our gain in specialization is also our loss, until perhaps only small-scale fishermen and their kind are left as the generalists who see things whole—and who will defend not their disciplines but our lives.

Coastal Star

*I*nside the bay, processing ships lie at anchor, cities unto themselves. At night, they are lit up like their own humming cosmos.

The *Coastal Star* is ours, the ship to which we transfer fish, the one that gives us orders. It took hours to pump off our first delivery, a training session for their new crew, and now, at midnight, the *M&M* is sent on a long rough-sea run far to the south.

Shelly and I, though, are trying to go in the other direction, north to her home. Done with my boat ride, I am headed for some shore time, and Shelly wants to wait out a little more weather. Neither of us wants a wild ride back out into seasick seas. I pack my duffle, she packs her pillow case, and we prepare to spend the night on the *Coastal Star.*

It is very space-age, getting aboard the ship; we're like the Jetson ladies in the old cartoon. From the ship's deck high above the *M&M*, a crane lowers a rope-sided capsule. Shelly and I unclip flotation vests from its side and put them on, then part lines and get into the cage. No sooner have we planted our feet and grabbed ahold, than we are whisked into the air, turning slowly in the wind and slashing rain.

Several stories above, we step out, remove the jackets, reclip them to the lines. A woman wearing a hardhat points us toward another space-age capsule—a molded plastic appendage that shelters a doorway. We make our way to it through a maze of huge vacuum hoses and bagged-up drift nets.

Inside, we come first to a smoke-filled balcony overlooking, through plastic sheeting, the processing line. The space is close, shoulder to shoulder with new hires who are, apparently, meant to be observing the procedures below. They stare at us as water runs off our raingear. Not unfriendly, not friendly. Mostly men but women, too—they are small and dark, most of them—Hispanic, Asian. They hold tight to their cigarettes. They are people who look, here at the end of the world, hungry to make a dollar. If independent fishermen are at the top of the fish-biz hierarchy, these factory workers are somewhere near the bottom, the unskilled labor that will make a fish into a product.

Below, on the line, salmon lie in messy piles—heads on, heads hanging—while knives hover over them, while water sprays. Manuals are left open on the stainless steel tables; workers cluster around for instruction. It is not a picture of efficiency. In fact, it doesn't look like anyone has much idea of what is to be done.

Shelly says, "Our fish." They may well be—salmon she snapped live and kicking into her skiff that morning. Definitely these came here aboard the *M&M*, cool in chilled sea water; definitely they rode the big fish-pump hose out of our hold. With luck, they will soon be relieved of their gills, guts, and gonads, not to mention their slime. They will be flash-frozen and sold to Japan, where discriminating shoppers will admire their redness of meat.

We find our way to the office, where a woman hands us bundles of bedding and shows us along narrow, brightly lit corridors, to a "dorm." It is a men's dorm, but there are only a couple of men assigned to it, at its far end.

We take bunks near the door. The room is large, but the tiered bunks fill it entirely, with only the narrowest of aisles between them. Privacy consists of curtains that close around each bunk; inside: one thin mattress, one hammock-style storage bag, one shelf, one tiny reading bulb.

I imagine the room during the season's peak, layered with tangled, exhausted bodies and not enough air. At least this hygienic ship doesn't hot-bunk; on other processors—some of those factory trawlers in the Bering Sea—as soon as one worker gets out of bed, someone on the opposite shift gets in.

I make my bed with clean sheets and the one light blanket. I drop into sleep without dreams, in the chamber of a throbbing heart.

SPARROW

he M&M *gets called back*, and it is still night when we are dropped by skiff at Stonewall Place. Rain hammers down, heavy and cold on my shoulders and hooded head, fogging my glasses. Behind Shelly, I stumble up the switchback trail toward her lighted house.

From out of the gray, three descending notes. A golden-crowned sparrow, awake, welcomes us, welcomes the day.

I have to smile, remembering all Burroughs's close attentions to the sparrows he met along his way. When the Harriman party stopped at Kodiak, the first thing Burroughs noted was a familiar song sparrow singing out from atop a weather vane. The bird was twice as large as his home bird, but its sweet-sweet trilling was recognizably the same. Burroughs, always looking for the familiar, could barely see the buildings, the lanes, the Russian faces for the sight of that one known bird puffing itself over a rooftop.

"Tsik-ez-lagh," this hidden bird sings to me. Those three drawn-out, descending notes. I used to hear them (because I read it in a book) as "three blind mice." Later, I learned that Alaska's discouraged homesteaders heard the song as "oh poor fool," and Keeler, of the Harriman expedition, reported that "some wag" they met along the way called the bird Weary Willy because its song sounded like "I'm so tired." These days, though, I hear the golden-crown's name in the language of the Dena'ina Athabaskans, the people native to where I live. They call it

Tsik'ezlagh, because this is what the bird says, its own syllabic name.

Burroughs, meeting the bird for the first time, weeks before he would try to trade for its skin, applied his careful ear and described the song as "strangely piercing, plaintive . . . very simple, but very appealing."

Tsik'ezlagh sings again as I trudge toward the light. Plaintive? I wouldn't call it that, nothing so depressive. Minor notes, yes, and simple, and appealing. Appealing to me because I know it so well. In this strange treeless land, I am as pleased as Burroughs ever was to be greeted in a language I recognize from home.

WHY THERE ARE NO TREES

*Why the timber should thus suddenly disappear on
the peninsula and islands is an open question.*

HENRY GANNETT,
"General Geography," *Harriman Alaska Series*

T*he ground around* my friends' home is so springy I feel as though each foot is pushed off by a thousand releasing coils, as though I barely set my feet to earth before they leap effortlessly back into the air. I have not felt this way since I was a child and used to imagine that if I could lift one foot before the other touched down, I could as well as fly.

Overhead, clouds blow past, and the sun, breaking between them, lays its own mosaic across the land, moving patches of dark and golden light all along the ridge and up the farther slopes.

It is early yet for the explosion of wildflowers that will come, though the lowest slopes are patched with baby pinks and blues and pale yellows. The little girls, Claire and Emma Teal, have been teaching me names: lupine, spring beauty, orchis, primrose, cinquefoil, anemone. A short climb higher the hillside is duller, the plants shorter, smaller, close to the ground, and wound through with the pale dead stems and grasses of last year. From a distance, it looks a plain carpet, and I have to bend low to make out just what a jungle it is—all that wiry crowberry leaf woven in among the larger, waxy bearberry leaf; the tiny pink blossoms of lingonberry; the smallest tundra willow with its red-pollen pussy willows; exotic leaves and stems and heathlike bells from plants I still can't begin to name. The whole tightly wrapped surface flutters like rippling water around me.

The wind. Yes, the wind. Everything here hugs the earth, or it will blow away.

DISTINCTIONS

My friends Shelly and Buck and their daughters have the distinction of living in the very last house on the continuous North American continent, an isolated homesite enhanced by hydropower, greenhouse, and general resourcefulness, a fair-weather skiff ride away from nearest human neighbors. Indeed, their small holding stands alone between water and the biggest backyard anyone could wish for—wildlife refuge that goes on and on, back up the peninsula. Outside of the scattered communities we passed on our way here, the Pacific coast of the Alaska Peninsula is entirely public conservation lands—Katmai National Park and Preserve, Becharof National Wildlife Refuge, Alaska Peninsula National Wildlife Refuge, Aniakchak National Monument and Preserve, Izembek National Wildlife Refuge.

The address here, Stonewall Place (no house number necessary), refers to the reef that extends out from their shore like rocks piled into a wall. At lowest tides they slip along it to gather *bidarkis*, gum boots, chitons—those plated mollusks pried from rocks and eaten raw with a dash of soy and wasabi.

From their guest cabin on the beach, I face Isanotski Strait, the pass that was later called False Pass by American mariners who thought it too treacherously shallow on its north side. Unimak Island, just across, rises pale green and truncated, its upper slopes and volcanic peaks cut off by clouds that look like the bottoms of aluminum cooking pots.

Before the first Russians came here, Unimak and the islands arcing for more than a thousand miles to its west were home to perhaps 16,000 prosperous Aleuts; Unimak, the largest of the islands, was also the most heavily populated. Enslavement, murder, relocation, and disease emptied the land, and by 1785, only 1,600 Aleuts were left. An 1849 survey counted a dozen remaining villages on Unimak, and the Harriman expedition would still have found several, had the *Elder* drawn close for a look. Unimak today is known for its wealth of prehistoric and historic sites, mostly unexplored—though pot digging is not unknown, treasure hunters being everywhere these days. Unimak is also known for its single town of False Pass, population 70.

DESERTION

Somewhere in their Alaska travels, the Harriman expeditioners were told of a "deserted" Indian village. Near the end of their return trip, they sought it out—the Tlingit village of Cape Fox, in southeast Alaska.

Burroughs: "There was a rumor that the Indians had nearly all died of smallpox a few years before, and that the few survivors had left under a superstitious fear, never to return. It was evident that the village had not been occupied for seven or eight years. Why not, therefore, secure some of these totem poles for the museums of the various colleges represented by members of the expedition? This was finally agreed upon, and all hands . . . fell to digging up and floating to the ship five or six of the more striking poles."

Our scribe is, himself, busy with other work. All day, in the shade of spruce trees, with Tlingit graves at his back, he writes up his notes. The others, the younger men and the ship's crew, take down the elaborately carved and painted poles—some of them sixty feet high and some of them, burial poles, containing ashes of the dead—and tow them out to the ship. Burroughs has no more comment upon this, instead recording the names of six birds, and that he picks salmonberries to eat with his lunch.

At some point during the day, a group photo is taken on the beach. A line of shake-roofed houses stands behind, with eight or ten totem poles before them. Just beyond the group, piled on the beach, are wooden screens, boxes, masks, carvings and other apparently ceremonial items obviously looted from the houses and ready to be hauled away.

From the graveyard, Harriman takes as souvenirs a pair of carved wooden bears, six feet tall. Someone else takes a Chilkoot blanket.

What do they think, these philanthropic men, who will make off with these private items, poles that memorialize ancestors and clans, religious and grave objects? I think I read in our narrator's words a polite questioning, a defensive posture. *Why not?*

At least one man among them knows why not. Muir, the only member to absent himself from the group photo on the beach, had twenty years earlier protested the "sacrilege" of collecting totem poles. He knows these objects belong not to individuals who give them up when they die or move away, but to cultures and history and spiritual life, and that robbing graveyards is wrong anywhere, always and forever.

The H.A.E. sacks the Tlingit heritage at Cape Fox as though the carvings and blankets are no more than mollusk shells and animal skins, just additional Alaska specimens to be studied or admired, as though they come from a wild place of replenishment.

There was a rumor . . . a convenient rumor, to be sure. Would it have made a difference if the H.A.E. had known that the Cape Fox Tlingits were living at Saxman, a baseless new town to which they had relocated with a missionary who insisted that everything they had formerly done to honor their ancestors and understand themselves, everything they had believed in and known for truth, was ignorant and unholy superstition?

Totem

I stare at the Edward Curtis photograph of a totem pole from Cape Fox, labeled "Beaver Totem, Deserted Village." It rises from an overgrowth of wide-bladed grasses and salmonberry brambles. Thick spruce trees crowd it from behind, dark background against which the aged wood of the cedar almost glows.

The totem figure is clearly a beaver. Paws clasp a stick, banded with dark and light coloring, that extends downward from either side of the mouth, where square, wood-cutting teeth close over it. There is a flattened nose, and large eyes, oval-shaped and nearly touching, with black irises, white in the corners. Another pair of paws folds over the head, around an ancient, pulpy, lichen-streaming top section I at first take for a tall forehead and then decide is the beaver's wide and flat tail.

Most striking of all, though, is the small human face that peers out from between the upper set of paws, above the beaver's eyes and in the center of its tail. It is that of a man, with deep, soulful eyes, and a downturned mouth. He looks out and down, at the photographer, at me. This small, humble person-face is not abstracted like the figure of the beaver, but appears in all its splendid detail as a real face of a real man.

I cannot help but make the comparison: this is the same face, the look, of a beleaguered Christ.

THE BIG SILENCE

From *Stonewall Place* I walk south along the beach under cloudy, swirling, ever-changing skies.

The narrowish beach slopes gently from the steep, hummocky hillsides and occasional rock cliffs on one side to the channel on the other, where the wind runs little williwaws over the surface. The walking is easy over sand and coarse gravel scattered with a few larger rocks. Here a fractured pink shell, there a tangle of weed. I collect two mussed eagle feathers and a geode agate with fine crystalline points. From the signal light at the south end of the pass I spot three or four of the big salmon-processing boats anchored in the bay below.

On the way back I climb the spongy hillside to look over an abandoned cabin, its door and windows open to the wind, the entry on one end collapsed to scrap. The wood is weathered gray, the tar paper shredding from the roof, and the back gable end tilts drunkenly away from the front. The grasses growing up around it whip in the wind. It is a small, impermanent habitation in a large, landscape—so much treeless country, water and sky, mountains and cloud, and wind for company. I can hear, as I could not hear until this moment, the thunderous silence of an absence.

The man who lived here back in the '40s, until sometime in the '50s, was a Norwegian known as Lonesome Einar. He lived alone and was, perhaps, not so much a hermit as a person who liked his space. Aside from his cabin, he had had a full workshop, with a forge, just above the beach, and a winch system for dragging up skiffs and logs. According to my friends, who have collected local history from the area's old-timers, Einar loved to walk and ski, and his trapping cabin was more than fifty miles back up the peninsula; one of the legends goes that he once found a log there he wanted, so he jogged back to this cabin to get his axe and then back to cut the log.

Was he truly lonesome, or just alone? My friends say that in his old age he got very strange and had to be taken away.

Was he strange first and sought his solitude, or did the solitude make him strange?

I try to imagine living alone in this vast open empty country. I know, though, to a person living here this land would not be empty at all, but full of all those nuances of life that only an intimate can recognize—the different sounds of different winds, the seasonal hatchings and bloomings, each fox, each shrew, each stick of wood. To live here alone a person would have to not be alone; he would need to hear the other voices, to be fully aware, to call upon his own imaginative powers.

Thoreau went off to his little cabin on Walden Pond. He was no hermit, though he sought in his experiment to find the quiet that might allow him "to live deliberately, to front only the essential facts of life." Thoreau would write, "I have a great deal of company in my house; especially in the morning, when nobody calls."

The gregarious John Muir lived for many years by himself in a humble Yosemite Valley shelter, and thought nothing of going off into the wild for days at a time with only a hunk of bread stuck into his pocket. When the Harriman expedition stopped at Muir Glacier in Glacier Bay, his little wood-scrap cabin lay in something of a shambles near the glacier's foot. Muir had hoped when he built the cabin years earlier to spend long stretches of solitary time there, listening to the groans and crackings of ice, but that dream never quite got realized.

Even John Burroughs, back on his farm, had his private places. First he had built himself a study away from his main house, and when that was not far enough for peace, he had moved well away, into the bark-covered cabin he called Slabsides. He identified with a great-uncle, who had lived in a hut in the woods and talked to himself. Great-uncle John, called by other relatives "a monstrous queer man," used to stand in the road and gaze all around him—something his great-nephew also did as a way of attending to his surroundings. As Burroughs got older, his thinking and writing turned more and more from observation to a worrying of philosophic questions; he collected his most mature writings under the title *Accepting the Universe*.

Lonesome Einar, then, falls well into a tradition of solitary thinking men. He did not commit the workings of his mind to paper—not that I know of—but I will assume that in his aloneness he concerned himself not just with the small practicalities of getting food and staying warm but with big questions: Who are we? What are we doing here? How do we stand in relationship to the rest, to nature or to what we might know as God?

I think it would be impossible to live anywhere in what remains of our big silences and not share this wonder.

To See What I Can See

With my friends' dog and my daypack loaded with camera, binoculars, jacket, lunch, and dog biscuit, I plan to hike behind the house, up the hillside, to the top of a ridge. It should take me about an hour to reach the ridge, and then I mean to just linger with the view, to see what I can see, and then come down again.

This is bear country, and I know to keep well away from the alder-choked gullies, out on the open tundra where I can see a long way. I hope to see a bear, actually. With my binoculars, I keep scanning the hillsides all around, far off and across Isanotski Strait to Unimak Island, hoping to find one grazing in the distance. But it is the middle of the day, not a prime time for bears to be out, and I know I'm not likely to spot one.

I make noise as I climb the hill—this too because of bears, to let any that may be nearby know I'm here. Although I sometimes arm myself with pepper spray for traveling along noisy creeks or when berry picking by myself, I mostly rely on making noise to protect myself from bears. Guns, I have always thought, only invite trouble. Bears that might simply have walked away all too often end up dead. And didn't I just hear of a man in False Pass who accidentally blew apart his own leg?

Today I call loudly to the dog, which races in circles, and I whistle and sing and sometimes just yell out, "Ho, ho, here I am!" The tundra, soft under my feet, retreats from early summer to late spring as I climb, from flowers to buds to the beginnings of new green under the alligator-skin bearberry leaves. I wish I could stay here until August, when this hillside will be heaped with mossberries, blueberries, and jewel-red lingonberries.

For now it is all promise. I feel it in my breath and in my step—that same drive, what is to come, the *becoming*. I could fairly float up the hill, as light as dust, as tumbleweed, as bird on wing. I *am* floating, gliding right on up. I am beyond sight of the house and up a broad slope, and then halfway up the hillside, steering myself up the middle and not forgetting to look back behind me at the water rushing through the pass and the mountains across and to the two draws on either side, and to the top ridge, always looking around me, taking it all in.

And then, at the edge of this large, still scene, my eye catches on the smallest sliver of movement. From the draw on my left, a hundred yards off, lifts a curve of brown. A bear's head. Its shoulders. Its chest. A bear rises out of the draw, all the way onto the crest of the hillside. It is, it seems, perfectly engaged in being a bear; it is watching the ground for a vole tunnel or a juicy patch of roots, and it hasn't noticed me, or the dog, at all.

I am amazed, disbelieving, even annoyed. Why is there a bear right there? Hasn't it heard all that noise I've been making? How can it be so obtuse not to see or smell me now?

I think I am not really afraid, though I am certainly very alert, measuring the bear, the space between us. It is a large bear, not the largest I've ever seen, but large, and very thick and rounded, very muscular, with seamless fur as deep as beds of moss.

"Ho, bear!" I call out and clap my hands. "Go away from here. You see me, I'm a person, you don't want to come this way. G'won now. Go away." I am pleased with my calmness, the fact that I am able to speak firmly to a bear, talk it away.

The bear looks at me and begins to run.

Toward me.

The rest happens very quickly.

I look for the dog, find it behind me, cowering.

I wave my arms over my head and at my sides. I yell again.

I register that the bear is a truly beautiful animal, with a handsome, dark-eyed face. Maybe it is not running so much as it is loping. It bounces over the tundra.

I give the dog a gentle kick. I have heard too many stories about dogs that chased after bears and then were chased back, delivering angry bears directly to their owners. I don't want that, but I think a couple of assertive barks would help.

Still yelling, still waving my arms, I back up slowly. The dog tries to get between my legs.

I am doing everything right. It is the bear that's breaking protocol. It is still coming toward me, undeterred. I have a sense that this bear doesn't suspect that there is anything in the world that it might need to be afraid of. It looks entirely comfortable with itself and with coming to look me over.

I wonder how much it will want to claw and chew on me before it leaves me alone. I wonder how bad the hurt has to be before I will prefer to be dead.

I remember a magazine photograph posted on the bulletin board at a fish hatchery where I worked many years ago, of a man who had been mauled by a bear. It was a horrific waxed purple face, with rearranged and some altogether missing features.

I wish the dog would step forward and sacrifice itself, and I wonder how terrible it will be to see a bear kill a dog, and whether I will be able to get away unnoticed while that happens.

I think very hard that I don't want to be eaten by a bear. I'm still shouting at the bear, still backing away slowly. My voice, I notice, is no longer calm. It is, in fact, quite high-pitched, almost shrill. I feel embarrassed about this. I'm not a person who panics. And I know not to scream; screaming sounds to a bear like a wounded animal, prey.

I know what to do. After the bear, as they say, "makes contact," I'll lie on the ground and pretend to be dead. But I have trouble imagining this. My every fiber tells me to stand and fight.

I think about pain. And about the people I love. I don't want them to have to imagine my death by bear attack. It would hurt them too much, to have to live with that.

The bear is just a few yards away. It hasn't so much as paused, though it is only moving now at a fast walk, as smooth as water. It is a perfect specimen of a bear, without a rough spot to its fur or the smallest scar anywhere on its face. I feel maddingly vulnerable. There is not a stick or stone for miles.

I'm yelling still. I hear my voice, which is not like my voice, shouting, "I'm nothing! I'm nothing!" I mean, of course, I'm nothing the bear wants to eat, nothing it should concern itself with.

When it happens, when I get knocked to the ground, I'll lie face down and try to protect my head and my belly.

As I think this, I think that when I'm down, my pack may help protect my back, and then I remember my pack and I sling it off in one quick motion and throw it in front of me, to one side of the bear. "Take that!" I yell. I remember the dog biscuit in it, and the almonds and orange, and I will the bear to rip apart my pack and eat everything in it while I walk away.

All I have left in my possession are the small binoculars in the pouch of my sweatshirt. I grip them in my fist and know my last effort, if the bear doesn't stop now, will be to throw the binoculars at its face.

The bear turns and drops its nose to the pack, and then it lifts its great head, and its eyes open very wide to show white all the way around—white with a crackling of broken blood vessels in the corners. Its look as it takes me in is one of undisguised, nostril-flaring horror, and then it twists itself around and leaps back the way it came. In its scurried retreat its stubby tail presses

so hard against its behind that the back end of the bear seems it
might overrun the front end. It never pauses, never looks back,
and then is gone, disappeared over the edge into the gully.

I breathe now and talk kindly to the trembling dog, spend a
minute collecting myself. My legs are slightly shaky. I have time
to think about the seven-year-old from a nearby village, who, a
few years ago, was not only killed but eaten by a bear. He had
been walking along a road with his mother and sister when a bear
stepped out of the brush. All three of them ran, and he was separated from the others, running through tall grass just like a caribou calf or anything else a bear might normally chase for its
food.

Buck and Shelly said that, afterward, people from all the villages around shot every bear they saw, with vengeance.

I look at my pack, lying on the open tundra, and then move
toward it, counting my paces. Five. The bear and I had been
twelve, maybe fifteen feet apart. I replay the whole thirty or forty
seconds of the encounter in my head, and it is clear to me that
the bear never really threatened me, wasn't charging me or
otherwise being aggressive. It was a young bear, and curious. It
didn't know what I was and wanted to find out, and the breeze,
blowing up the mountain, never carried my scent to it. Not until
it put its nose to my pack did it realize that there was something
very unusual about me, and then it did exactly what it was sup-
posed to do.

Muir wrote that, in Alaska, bears move "as if the country had
belonged to them always." This country, at least much of it, still
does belong to them. Compared to the 31,000 brown bears that
inhabit modern-day Alaska, fewer than a thousand still live in all
the lower forty-eight states, and those survive outside of parks
only by timidly skulking around under the cover of darkness and
thick brush. In most of the world there is not a chance of living
with bears or witnessing much of anything beyond the bossy do-
minion of humankind.

What is the world if it isn't more than ourselves? There is something to be said for keeping alive mystery, for feeling both wonder and submissiveness.

I take out my binoculars and scan all around again. The hill-sides are vast and still and maybe even a little more lovely than before, the colors sharper. The tundra has a spikey, peppery smell I had not noticed earlier. Clouds like flannel sheets hang decorously on the mountains across the way. In the pass, a barge powers through against the tide; its name, I make out through the glasses, is *Confidence*.

A drop of rain hits lightly on my cheek. I take it for my cue. In minutes I have got the house in sight, and then I am beside the creek and waterpipe and by the magpies' nest and on the porch, and then I am safely inside four walls, pouring myself a cup of hot tea and lifting it with two hands to my dry lips.

WILDERNESS

They *met wilderness* in Howling Valley.

It was Muir's suggestion to Harriman, that a valley he knew inland from Glacier Bay, through a pass just beyond Muir Glacier, would be just the place to look for big game, including the elusive bear. In "Howling Valley" they would hear the echoing cries of hundreds of wolves and find all manner of animals, large and small.

Muir neglected to mention that in his own, earlier travels to the valley, he'd nearly met disaster. Once, twenty years before, with his faithful dog Stikeen, he had gotten lost negotiating the glacier's knife-edge crevasses and had to be rescued by his Indian guides. Another time, he apparently met with a pack of snapping wolves and may or may not have had to fight them off with his hiking staff.

Harriman, Merriam, Grinnell, and the rest of a hunting party, including packers and an old scout who had once worked for Custer, rushed off in search of bear. For eighteen miles they trekked over rocky morraine and crevassed glacier, through pouring rain, across snow fields where they sank to their waists. They struggled all night because the glacier proved too cold to bed upon. They struggled past the point where one of the wiser packers turned back; the rest were driven on by Harriman's own doggedness. Finally they reached the top of the pass and looked down into a snow-buried, trackless, and silent valley.

The comforts of the ship were far behind. The men were surrounded by wilderness still locked in winter, dangerous and unforgiving. I imagine a chill passing through them, through even the experienced packers and old scout. There was no romance to be found in the obscurity of white, in the cold. A person didn't stroll about here and admire the views; a person stood in awe, had to feel his smallness, his insignificance. He had to know there was something greater than himself, beyond all his control.

Muir, who had stayed behind to escort the Harriman women upon his glacier, was perhaps at the moment climbing aboard an iceberg to pace off its 700-foot length or sharing with some Indians a meal of gull eggs and wild celery dipped in seal oil. ("The petioles were hollow but crisp, and tasted well.") He had—knowingly or unknowingly—sent the hunters on a fool's errand, but he understood this about wild places: that their values lay not in what could be conquered but in the humility they forced upon man. He knew we needed such places, would always need them, not as warehouses of goods but as temples for our souls.

The party turned back. Sleepless, blistered, exhausted, and half-frozen, they trudged for eighteen more misery-filled miles, back to the shore.

Where Burroughs, greeting them with sympathy, would scratch into his notes the thought that all the howling might belong to Muir's imagination.

Apparitions

Our hunters still dreamed of bears.

Burroughs,
"Narrative of the Expedition"

Sometimes, *it was rumored*, polar bears could be found on St. Lawrence Island, in the Bering Sea. They rode the sea ice down out of the north, and then were stranded there for the summer, until the pack ice paved their way back.

This possibility took Harriman ashore in fog and rain, rifle in hand. What a bonus it might be: a white bear to match his brown, the pair to surround his doorway like bookends. Dark and light. Pepper and salt. The exotic and the more truly exotic.

The Harriman daughters, tagging along with the scientists to collect yet more birds, spotted them first.

"Bears!"

It is Merriam who goes after them, tracking them through the parting, now enclosing, fog. Their ghostly, swaying backs are just visible over a rise toward the water. He pauses once, fumbling to change his shotgun shells. The two backs stroll slowly, meanderingly, and he creeps after them. He knows he will get only one shot, and he wants it to be a killing one. One mile, then another. He loses them behind swirling, fibrous clouds, then catches them in misty sight along a cut-away shore. He is close enough, almost, for a shot. He raises his gun.

The cackle is so unexpected, so jarring, he thinks in that instant that his pinnings to the known world have shaken loose, and then the long necks reach up and around, the pair of swans glare at him from coal-point eyes. One stretches out its snapping wings, scuttles closer to the other's side. Merriam can only lower his gun and stare, adjusting his eyes like the controls on a telescope, drawing back from the farthest deep-space view to near-distance, from the Milky Way to the moon. The ridge lies only paces away, a mere rise in the tundra. The distant boulder becomes a foreground rock. Those far and furry backs stand just there, metamorphosed into mounds of feathers, elegantly avian. The two swans slide their necks around, chortle again in warning, raise heavy wings that shine almost luminous in the light.

Merriam focuses on individual wing feathers. He sees the gray space in the opening of one bird's beak, and the tip of the bird's tongue against it. Everything he knows to be real has fallen into new space, a corrected and most absorbing dimension.

Later, he will tell the story with good humor, less embarrassed, I think, than impressed by what can seem to be. The others will label the occasion Merriam's wild-goose chase. The souvenir photo album given to every participant after the trip will include one oddly labeled photo—a pair of fuzzy baby geese running loose on the ship's deck, titled "Polar Bears, St. Lawrence Island."

If a bird can be a bear—and surely it can be, as a porcupine or a tree stump can sometimes be a bear, a field can be a sea, a firefly can be a star, and a crack in a wall can be the face of God's mother—what can we know to be truly fixed in this world?

The question is rhetorical, but I find something of an answer in Merriam's stalking, in Burroughs's narrow attentiveness, in my own failures in sight and anticipation. Perception, like beauty, lies in the beholder's eye, and may, it seems, have more to do with expectations than with anything resembling a clear and scientific truth. If what we see is governed by what we look for and how we go about looking, then the possibilities are as small or large as our very vision.

First Fish

When Buck returns from the opening, he has with him the year's first fish, the first red salmon he had pulled from a net two days before and set apart from the catch he had sold. It is still shimmery cold-ocean blue, its scales like tiny metallic plates.

Virtually all the Native peoples of the North Pacific, who depended for their lives on the returns of spawning salmon, honored the season's first-caught fish with some sort of ceremony. Buck, who grew up nowhere near ocean or salmon but in America's great grainy Midwest, is determined to continue this tradition. He and his daughters set off down the beach to collect a symbolic mussel shell.

The mussel shell is important because such shells, sharpened against sandstone, were used by Canada's Nootka Indians for salmon-cutting knives, even, apparently, after metal knives became available. The traditional Nootka thought that only shell showed the proper respect to salmon.

The Nootka live a couple of thousand coastal miles from here, in a home shared with the large sea mussel, *Mytilus californianus*. The Aleut ancestors who lived on this end of land and on the long sweep of islands to the west are unlikely to have relied on their smaller and fewer mussel shells in the same way, but they almost certainly had their own ways of celebrating the salmon's return. Those ways, like so much of the ancient wisdom, were long ago overrun by Russian Orthodoxy and Western beliefs; they were hidden, adulterated, adapted, and lost.

Buck has no one to tell him how the original people of this place welcomed salmon, and he instead finds what he can in books about other cultures, and in what the place itself suggests. In their own way, he and his family are as dependent on salmon as any who have lived here, and he wants Claire and Emma Teal to know this, to understand connections and responsibility and what it is to honor the creatures that feed them.

When I see him and the girls again, they have cut the salmon open with a knife and laid it on a loosely woven grass mat atop a wooden tray—the two fillets, the belly, the skeins of eggs. The meat is deep red and glossy, the eggs like jewels in transparent purses.

The girls pluck soft, downy fluff from near the bases of several eagle feathers and sprinkle it onto the fish, and then onto their hair, and their father's hair, and onto the rest of us who have gathered there. Like cutting with a mussel shell, the sprinkling of down belongs to Northwest Indian cultures. Eagle down symbolizes peace and friendship, and in ceremonial use it welcomes salmon and brings good luck to all who share in the preparation and eating. I bend to get my blessing, gladly.

Buck makes a wood fire in the barbecue and cooks the fillets. He and the girls carry the carcass—head and tail, bones and innards—down to the beach and throw it back to the sea. The fish's remains will wash back to the land of the Salmon People and will answer to the others there, that, yes, this first fish was treated with honor and respect. The other salmon then will freely follow, giving themselves, too, to be caught and eaten.

The fish is cooked, set on a platter. We crowd around the table—Buck's family, crew, myself. There is a steaming pot of rice and a salad of greenhouse greens and fireweed shoots.

Buck reads solemnly from a book about Indian fishing traditions, about the First Salmon Ceremony, that time of joy and renewal. The words tell us that the ceremony reminded people of nature's rhythmic cycles and the interdependence of all beings, that it was a time to perpetuate ancient customs and reinforce regulating taboos and a time of thankfulness—for making it through to another year and for the return of life-giving salmon. The words explain, but this I know: the good smell of fatty wood-cooked salmon, the way it will flake along the muscle grain, the melting taste. I believe in the land of the Salmon People no more and no less than I believe in a heaven, but I know how salmon will feed me, and how the carcass in the sea will feed others, how the cormorant and the amphipod and the sea lion all need one another, how central salmon is to us all, and where responsibility begins.

Emma Teal, with down in her hair, licks her bottom lip.

Buck reads a Kwakiutl prayer:

> *Welcome, friend Swimmer,*
> *we have met again in good health.*
> *Welcome, Supernatural One,*
> *you, Long-Life Maker,*
> *for you come to set me right again*
> *as is always done by you.*

For another minute we are, all of us, silently, thoughtfully, thankful, and then—like people in salmon country forever before us—we feast.

Words Have Meaning

It took many years for Mr. Duncan to change these Indians from the wild men that they were when he first met them, to the respectable and civilized people that they now are.

GEORGE BIRD GRINNELL,
"The Natives of the Alaska Coast Region,"
Harriman Alaska Series

At Metlakatla, back in southeast Alaska, the Harriman expeditioners attended a Sunday service in which the missionary William Duncan preached to his rapt flock entirely in the Tsimshian language.

"A vague, gutteral, featureless sort of language," Burroughs calls it. I see him watching Duncan's gestures and studying the attentive parishioners, who never so much as turn to look at the visitors in the back. The Tsimshians appear to him, in their suits and dresses, the women under bright silk scarves or extremely fashionable hats, just the same as any church-going people in any village in New York or New England. Their two-towered church, the largest in all Alaska, seats eight hundred in carved and polished pews. Today it is nearly full. Burroughs tips back his head to admire the vaulting ceiling and the native woods—the glowing yellow cedar—with which the building was so skillfully constructed. When the organ plays and the people sing—old familiar Anglican hymns with Tsimshian words—he is impressed in a major way; the voices and music are as well trained and harmonious as any he would find back in his rural East.

Our friend surely finds as surprising as anything in Metlakatla the strange sounds flowing from the mouth of a white man, a Scot and a Christian. Although Burroughs isn't much of a churchgoer himself, he must know how unusual this is. Few missionaries have ever embraced the language of their converts so willingly and with such success; to do so would seem to them as "backward" as adopting Native mythology or dance or potlatch traditions. Is not language the primary, most significant expression of a culture, and was it not those "inferior" cultures that missionaries desired to replace with a "superior," Christian, civilized one? Besides, everyone knew that Native languages were simply incapable of expressing Christian thought. Only by compelling Natives to speak English and only English was there any chance of making good Christians out of them.

But Duncan, this morning, has already told the expeditioners his story, and perhaps Burroughs now thinks over these facts: that the missionary began studying Tsimshian almost the day he arrived on the Pacific Coast, spending four hours a day with a tutor who knew very little English himself, so that they, really, were teaching one another, learning together. In his first month Duncan memorized 1,500 words, and within eight months he was formally addressing the people, articulating complex spiritual concepts about why he had come to live among them and what he hoped for their future.

Duncan might have told the expeditioners what he has told others—that he took to Tsimshian with perfect naturalness, that he felt more at ease in that language than in English.

Burroughs—who never had an ear for languages himself, unless they were those of birds—might marvel at this, but it is unlikely to occur to him that in another couple of generations the Tsimshian tongue will no longer be passed from parent to child, and that a century hence the only Metlakatlan speakers of Tsimshian will be elderly. He cannot know that Alaska Tsimshian will become one more casualty of American assimilation policy that for fifty years will absolutely forbid Native languages from being spoken in schools. Children who do so will be beaten.

The loss of the Tsimshian language will mirror the loss of other Alaska languages, other Native American languages, the languages of the world.

In his pew Burroughs hears what sounds to him vague, gutteral, and featureless only because he is not of the place that owns them. The sounds and the words, the language structure and precision, belong to their part of the world as surely as any cedar tree, any blue mussel, any of the birds he goes about listening for so eagerly. Tsimshian is as organic as the soil, and it grew as anything living grows, as a part of something ancient, complex, and highly adapted. It holds all the words and all the concepts needed by its people to live right and well.

When the Tsimshian language dies, it will be extinct, as dead and gone forever as any last shot-down or poisoned bird.

Burroughs sits upright in his pew and hears the people sing, and he cannot imagine the silence that will come, and that all the world will be less sung, and simply less, diminished for us all, without the Tsimshian voice, without the full chorus of naming and knowing.

Traffic

Shelly and I stand at the shoreline, at her fish-cleaning table, and work on salmon—three kings and a dozen reds. I slime, running a dull knife over silvery sides until the dark, bacteria-rich ooze stops working out of the skin. Shelly heads and guts a few reds for canning, and the rest we fillet and set in a tote of brine for later smoking. We save heads and roe for her friend Stanley, who favors both.

The tide pours in through the pass, like a river. A flock of sitting seagulls sweeps by on the current. The water is bruise-colored, darker than the slate clouds that hang low over Unimak. Vessel traffic is continuous—a squat yellow barge, with *pilot* lettered across the bow; local boats heading down to Ikatan Bay; drift boats and tenders aimed for the Port Moller side, for Bristol Bay. A sea lion surfs past, close to our shore, and an immature eagle tilts above us, casting its eye upon our gleaming fish. A helicopter clatters low along the channel, toward the village of False Pass, speeding someone important, or suffering, from processing ship to out-going plane.

A new sound comes to us across the water, and we look up to see a tall column of spray falling back to sea. And another. Two whales—grays, Shelly says—are coming through the pass, well along on their migrations. But these two are not simply traveling, surfacing for breath and rolling on through. One and then the other raises its whole huge head from the water and slaps down its chin. They leap out half their godly lengths, thirty or forty tons crashing again to sea. Diving, they smack their wide, shapely tail flukes and then surface sideways, throwing flippers into the air like the arched arms of only slightly awkward ballerinas.

Shelly and I set down our knives. We watch the leaping, the shaking, the crashing, the blowing, the slapping and flapping. These whales want to be seen, apparently. Or they want to see. They lift their heads in what whale watchers call "spyhopping," thought perhaps to be their way of looking around for landmarks by which to place themselves in the upright, supersurface world. Perhaps there is something about being in the narrow pass that excites them. Perhaps it is all the other traffic, and they are acting as their own foghorns and lights, letting everyone know they are coming through. Perhaps they are performing for one another. They are cavorting, whale-style.

We humans think we know so much, but we can only guess at why these enormous creatures do the things they do. Our species was clever enough to hunt gray whales to within a few thousand of being no more, but we have also been smart enough to protect them, and today perhaps 20,000 swim the eastern Pacific. Most of them take a wider turn through Unimak Pass, but some hundreds must take this Isanotski shortcut.

Shelly tells me about the dead gray, earlier that spring, that washed up in the pass. Judging from its appearance, it had been killed and partially eaten by killer whales, predators known to surround much larger whales and attack like wolves do a moose. The carcass was towed into False Pass, where the villagers, scavengers themselves, stripped it of its fat, which they call in Aleut *alla-x̂*.

Boats are passing the whales. One open skiff cuts its engine and drifts between them, a small aluminum shell not much longer than the whales' tails are wide. Shelly and I stand at the corner of the continent, beside the thoroughfare that connects everything to the south to everything to the north. The parade goes on—the quieting whales now beyond the reef, the boats, the salmon swimming through the pass.

Tub

The men have gone fishing, and the ladies of the place—Shelly, her sister-in-law, the two little girls, and I—take over the hot tub at the edge of the sea. We shed our sweatpants, our turtlenecks and pile jackets, our wool socks and underclothes, and ease ourselves into the big wooden tub, its water heated by a submerged woodstove.

The warmth is penetrating, exquisite.

I sink to my chin, feel the ends of my hair dampen against my neck. Magpies pick at something on the beach. The water before us, gray and wind-ruffled, slips past with the tide. Across the strait, the mountaintops are lost in swirly opaque cloud, but the lower hillsides where the snow gulleys melt into green valley are bathed in soft evening light, so clear it is like looking through liquid. "Lucid light," Burroughs called this, and the green his "carpet of verdure."

We soak ourselves into torpidness. The clouds overhead shred into banners with pink and lemon edges. We lift ourselves to cool, sink to be held by the heat, to be massaged by the silky water we stir. The little girls, busy as ever, scoot around the tub, making waves, pouring cups of water over their heads. Now they climb out and run to the beach, wade past their knees in the frigid Aleutian water. They are as beautiful as any creatures anywhere, these pale, utterly at-ease inhabitants of this farthest place. I can, at this moment, think of no more pleasing scene in all the world—Claire and Emma Teal there in the light, with the water and the green and snow-streaked mountains, the great cottony sky all around.

Poor John Burroughs. He shivered through his entire Alaska voyage, bundled in his heaviest clothes and never warm enough. I have no idea what bathing facilities were aboard the *Elder*, but surely the ship's coal fires heated water for the occasional Victorian bath. Would that he could imagine us, five naked females in a steaming tub, now rising to cool and to watch the rough-legged hawks at the cliff chase away a raven, all of us warmed through to our bones, perfectly content.

SCENERY

Late in their expedition, returning south, the *Elder* steams through rock-and-roll seas under some of the clearest skies and longest views the Harriman party have experienced in all their two months and thousands of miles. For two days the snow peaks of the Fairweather Range cut like glowing stencils into blinding blue sky.

Muir marks the scene as the "most glorious" of the entire trip, while the less effusive Burroughs writes that they are treated to such a view "as is seldom granted to voyagers."

One of the group's artists manages with freezing hands to paint Mount Saint Elias at sunrise, before the seas pick up, but the second, later-rising artist meets only frustration.

Burroughs, in his stateroom—seasick again, I'm quite sure—notes when they pass Mount Fairweather late in the day that the light reflected from the mountain floods into his room "like that of an enormous full moon." The whole day was for him a day of blue and white—sea, sky, and mountains. I imagine Burroughs at his round window, holding his eyes to the steady mountain edges as the ship pitches and rolls. Surely he knows this trick for fighting motion sickness. This, and letting the wind beat fresh air into you while you squeeze the rail and spit into the sea.

The previous day he exulted in the baskets of wild strawberries brought by Indians in Yakutat Bay. Always comparing to what he knew and idealized back in the East, he found the berries at first puny and pale, "uninviting," and only after trying them admitted to a "really excellent" flavor. Lying in his bunk with an eyeful of blue and white, does he think of these berries, or of his Concord grapes heavy on the vines at home? Or does he banish all thoughts of food and only long for soonest arrival on level ground, for the comfort of *terra firma*?

And so, as the ship slowly slides past the Fairweather Range, expedition members gather one more time on deck to admire what is almost the end of their Alaska scenery. The younger Harriman children are perhaps again dressed in their sailor suits. And where is the expedition's host? Mr. and Mrs. Harriman are sitting in deck chairs, covered with robes, on the opposite side of the ship, facing the long blue horizon of the North Pacific Ocean. There isn't a mountain, or any land at all, in their sight.

Merriam hurries around the deck to find them. He calls out, "You're missing the most glorious scenery of the whole trip."

Harriman stays where he is. The return trip has been long for him, and he is impatient to get back to his manic-paced work life. He tells Merriam, "I don't give a damn if I never see any more scenery!"

Mrs. Harriman's reaction is unrecorded. Has she had enough scenery now, too? Does she simply smile at Merriam, excusing her husband's bad behavior? Or does she toss aside her blanket and tug down her skirt, and go off to see that last, most spectacular view? I choose the latter for her—for this nineteenth-century woman about whom we know so little, who is always in the background. I want her to have that last pleasure—that "most glorious" coastline under clear skies. I want her to have that gladness for herself, and to know with me that there is such beauty in this world.

The Clean Industry

[Alaska's] grandeur is more valuable than the gold or the fish or the timber, for it will never be exhausted. This value, measured by direct returns in money received from tourists, will be enormous.

Gannett,
"General Geography"

I f *geographer Gannett* were alive today, he might instead work for the Alaska Chamber of Commerce or the Anchorage Visitors Bureau. He had the right attitude for the job.

And he wasn't far off. Tourism today employs more people in Alaska than any industry except fishing, and it is the third largest contributor (after oil and fishing) to the state's economy.

The incipient industry then brought people to gawk at mountains and glaciers and to buy cedarbark baskets from the Tlingit women who spread them on the sidewalks in Sitka and Juneau. In Muir's beloved Glacier Bay, the expedition found plank walks laid along the shoreline, the better to protect the shoes of visiting cruise passengers. Scenery was the draw—those big, romantic, unspoiled vistas, of which Curtis shot 5,000 splendid photos.

When it came to praising tourism, Muir was the expedition's dissenter. He didn't like tourists; he thought them disrespectful of the land he so loved, trivial in their interests. In his earlier trips to Glacier Bay he had noticed how quickly the cruise passengers stopped their gazing whenever the dinner bell rang.

Today the number of tour boats allowed in Glacier Bay is limited, to protect the humpback whales that graze there and the "quality of the visitor experience." The views throughout much of southeast Alaska are of clear-cut forests. Farther north, tourists pack the hotels and buses at Denali and airplanes buzz like flies around the continent's highest peak. The popular salmon rivers are shoulder-to-shoulder with fishermen, hooking one another in what's become known as "combat fishing." The roads are pokey with huge motorhomes. *Pan gold* (from seeded streams) *here! Ride a dogsled* (on wheels) *here! Take a tramway to the top!*

A woman who works as a naturalist at Katmai National Park told me that when a brown bear stopped in the middle of a trail to nurse its twin cubs, the tourists—once they had snapped their photos—threatened her with bodily harm if she didn't hurry up and make the bear move out of the way. It was almost cocktail hour back at their lodgings, and they didn't want to be late.

I have long belonged to John Muir's Sierra Club and to the Alaska Marine Conservation Council, but recently I joined a new, local, half-in-fun group—Folks Against Rampant Tourism.

FALSE PASS

I *arrive in the village* by skiff to catch a plane that will carry me to Cold Bay, where I will board a jet for Anchorage. I know how far I am from home when my one-way ticket sets me back $538.

False Pass is quiet, its residents off fishing or otherwise engaged in making a summer's living. A few children run along the boardwalks as I lug my belongings through town, past the Peter Pan stockroom, the small company houses, all vestiges of the cannery around which the town grew up in 1917. Though the cannery building itself burned down fifteen years ago, Peter Pan still provides the fishing fleet with fuel, supplies, and gear storage.

Then I'm at Per's, which is where I am to wait for my plane. Two men, hats pulled down over their eyes, are sacked out in chairs; a large-screen television is tuned to CNN News. Per, a tierce-chested, accented Swede, tears around the building and the town, rounding up passengers. There are two planes, a charter and a scheduled one, and it is his apparently very complicated job to get the right people and the outgoing mail to the right planes. He calls Cold Bay on the phone to discuss the time, rouses the sleeping men, and agonizes over the location of a Filipino woman from one of the processing ships who went off to the store for cigarettes and hasn't returned. He tells me the complete and intimate histories of two of my fellow passengers, who have been "drinking for days and fighting with everyone." Per volunteers that he likes life in False Pass because "there's always something to do." His twin daughters are in college studying baking and auto mechanics, respectively, and his son, a diesel mechanic, is in the hospital because he accidentally shot himself in the leg. He told them, all three of his children, that they had to learn something besides fishing. This year the fishing is as poor as he can remember.

At the airstrip, a woman quizzes me (Who are you? Where are you from? Why are you here? Do you like it here?), then directs me to watch over her departing visitor, a young girl I am to deliver to a certain woman in Cold Bay and no one else. More people arrive for the plane. There won't be room for everyone, and the couple with a bottle in a paper bag cheerfully unload one of their cases of beer and stay behind.

We fly off under clouds, leaving False Pass, leaving Unimak Island. Just once in my week on the strait did I glimpse the top of an island mountain, the one called Roundtop, never the higher volcanic peaks that so delighted Burroughs he likened them to Mount Fuji.

After everything, here is this one small, shrinking village, inhabited by people partly Aleut, partly other, a people born to wind and fish or drawn from the eight corners of the world to this version of the good life. The village at my last look is a few shining roofs, fuel tanks, pale winding roads across the close-cropped tundra. My eye lingers on the new, absurdly substantial ferry dock. Because of the dock, now the state ferry comes to False Pass. It stops once a month in summer, on its way out the Aleutian Chain. It doesn't stop on the return trip.

It is a thought I like: there is no going home from here. Come this way west and then, imagine there are no planes, imagine the *M&M* will not return to the Homer harbor at the end of the season. Imagine there is only this far country, and no turning around.

WASHINGTON, D.C.

QUEST

Months *later*, in the Smithsonian Natural History Museum, I wander among holiday crowds of children, looking for totem poles. On the ground floor there is a pair identified as Haida. Upstairs, in the dusty, utterly lifeless Native Cultures hall—where the full-scale models and fakey faces of what are meant to be *people* give me a major case of the creeps— I find one labeled "probably Tlingit, Southeast Alaska." It clearly was a house pole, forming a portal through a giant bear's mouth at the base; the bear is topped by an orca and, above that, two watchman figures looking out. There is no entering anywhere now, though; its whole lower portion, to a height beyond anyone's reach, is protected behind a Plexiglas shield.

"Probably Tlingit." This is far too weak an identification for something taken by the H.A.E. from Cape Fox.

I call the Smithsonian anthropology department, where a very nice woman offers to research their totem poles and find out if they have any collected by the Harriman expedition. When I call her back, she tells me they don't—only some small argillite models that Mrs. Harriman purchased from Governor John Brady at the Sitka stop.

Two poles went to the Chicago Field Museum, two to the Museum of the California Academy of Sciences, one to the University of Michigan. One, I thought, would surely have gone to our national museum, especially since three of the principal

H.A.E. scientists were associated with the Smithsonian. But apparently this was not the case . . .

I feel oddly dissatisfied. I wanted to put my own eyes on a Cape Fox pole, to see it lifelike, full-scale, in actual wood and paint. I wanted to be able to imagine it back in its place. I wanted, I think, to apologize. But I wanted, too, to look at other people looking at it, and to think that there was value in having it here, that it *means* something to our larger national picture—a meaning not of theft but of beauty and spiritual life, of some kind of understanding. I wanted to be convinced that there was *some* justification for its removal from its rainy shore, where it would by now be moldering back to earth.

BONES

I *browse through* the Smithsonian's collection of skeletons, admiring the artful articulation of pelvises and thigh bones, vertebrae and rib cages, all that mysterious machinery to ostrich, to bat, to gazelle.

The three sirenians hang together on one large wall. Manatee, dugong, and Steller's sea cow. One from Florida and the West Indies, endangered. One from the Indian Ocean and considered rare. One formerly from the North Pacific, extinct.

Always, I have known of the Steller's sea cow, known that Vitus Bering's onboard naturalist, on the return leg of the first Russian voyage to Alaska, had barely described it before it was gone for all time. I have held the animal like myth in my imagination, believing in it as one believes the inhabitants of a bestiary, wanting it to have its place and its meaning. I have carried a mental picture of it—big and leathery, slow-paddling, chewing kelp like a cow—through my life.

But I had no idea!

The sea cow skeleton on the wall is a giant, at least thirty feet long, much larger than I had ever guessed, dwarfing the manatee and the dugong below it. Its ribs are thick and massive, its skull beaked. Its stumpy front limbs hang as though silently paddling, and its tail stretches far behind like a trailing flag.

The placard beside the exhibit tells me the sirenians (so named because ancient Greek sailors apparently mistook an upright, nursing dugong for a mermaid) are related not to whales but to elephants. They are strictly herbivorous, which accounts for the excellent taste of their meat.

I know more. The fossil record reveals sea cows, 100,000 years ago, living all around the North Pacific coast, from California to Japan. By the time Georg Steller first described them, in 1741, they were confined to the inshore areas of two uninhabited islands—the last two islands of the chain that runs from Alaska to the Russian coast. They numbered, perhaps, in the thousands.

Bering's ship wrecked on one of those islands. The sea cows did not flee when hunted, and their meat and fat, described as better than beef and butter, kept men alive. They were so docile Steller stroked them from shore at high tide. He took meticulous note of their appearance and behavior and paid some of the sailors in tobacco to help him measure and dissect one specimen.

After the sea cow's discovery, ships sailing from Russia to Alaska in the fur trade routinely stopped on the islands to provision with meat, and other fur hunters made a habit of wintering there. One sea cow was said to feed thirty-three men for a month.

Within twenty-seven years, the Steller's sea cow was gone.

No Steller's sea cow was ever collected for science, and the careful drawings made by Steller's assistant disappeared on their way across Siberia. This museum skeleton is a composite, made up of bones gathered on the beaches of Bering Island long after the live animal ceased to exist, and pieced together by guess.

I try to picture the bundle of bones in its flesh, thousands of pounds of it, all the fat it needed to keep warm. Its thick hide was said to be so furrowed it looked like the bark of an old oak tree. The tail flukes, missing from the skeleton, were horizontal, like a whale's. The forelimbs were flattened flippers, with soles set with bristles, like scrub brushes, to help the grazing animal hold its position in the rocky shallows. The fleshed head, Steller said, looked very much like a buffalo's, especially in its lips. Instead of teeth, it had two boney plates it ground kelp leaves between. The eyes were the size of a sheep's, and the ears couldn't be seen at all. The poorly witted sea cow was equipped neither to fight nor to flee, only to be a constant eating machine.

Twenty years before the Harriman Alaska Expedition, a visitor to Bering Island reported that local people claimed they had seen the Steller's sea cow. Twenty years before my own exploration to the end of the Alaska Peninsula, rumors of sea cow sightings still floated from Russia, and there was talk of mounting an expedition. For all I know, there is talk still, and powerful belief. I want to believe, myself, that some stealth population, around some small island somewhere at the edge of the world, survives.

And there is Burroughs, on the edge of my consciousness, trying still to bring his vision up to Alaska's scale. I don't know whether he knew of the extinct Steller's sea cow, though I wouldn't be surprised if Dall or one of the other H.A.E. scientists had lectured about Bering's expedition and the extraordinary work of Steller. My guess is that, if Burroughs ever—in his or my wildest dreams—had been able to see a living sea cow, what he would surely have first noticed were the flocks of ordinary seagulls that fed on the animal's great barky back as it grazed in the shallows, just as field birds pick bugs from the backs of cattle.

On the Street

The sidewalk overflows with foreign tourists, well-dressed Washington office workers, students, an aged woman tossing popcorn to swarms of pigeons and casual starlings. (I wonder: what would Burroughs think—these brazen birds under our feet not American at all, but invasive species from Europe? Would he be disappointed with what, bird-wise, we have become?) I make my way through, turning my shoulders, holding tight to my Metro card and absorbing the ambiance, the historical presence and anonymity of my country's capital.

This was Burroughs's city, back when he was a young man trying to support a wife with a job as a government clerk and beginning to find his writer's voice. He had lived here in a frightening and amazing time—during the Civil War, and when Alaska belonged to the Russians and when it was bought by America. He had guarded millions of bank notes and taken long walks with his good friend Walt. It was Whitman himself who confused readers for all time by suggesting the title of Burroughs's first essay collection, *Wake-Robin*—the common name of a spring flower, the white trillium, and nothing to do with the exhortation of a red-breasted bird.

Surely, living here, in the Washington that was little more than a town set amid swamps and woods, Burroughs never for a moment considered traveling to Alaska, never imagined the need or desire. The landscape all around was big enough, more than enough, and all he ever really wanted was to get back to the pastoral purity of his Catskills boyhood. That attitude was not a character flaw, I think, though Muir seemed to find it so. Muir was the extraordinary one—truly the seer that Burroughs suspected, a man out of time. Burroughs belonged very well to his era; his books, otherwise, would never have had such wide appeal.

All these people, all around me—who among them has ever even heard of John Burroughs? Of Edward H. Harriman, of whom, when he died just ten years after his Alaska expedition (his Type A personality having apparently worn itself out), Burroughs would say, he was a great man, "as great as Napoleon"?

Who in this crowd would know the first thing about the Harriman Alaska Expedition, or have a reason to wonder about it? Would that man in the orange tie, or that frowning young woman care about the five thousand mounted insect specimens? The thirteen thick volumes of report that Merriam slaved over with a whole flock of additional scientists and assistants, for twelve years (never completing the volumes on mammals, of which he was primarily in charge)? Who here would be intrigued by the elaborate study of sea stars and their sensory mechanisms, backed up by Darwinian theory? The documentary photographs and artwork, and the sound recordings of singing and trumpet-playing Tlingits? The complicated pioneering analysis of glacial behavior?

The scientific results of the Harriman expedition, historians tell us, *were* significant. Many volumes of the report became reference works for later researchers.

But my concern is not with results. It was enough for me that it happened, that Harriman and Merriam, Burroughs and Muir, Grinnell and Curtis and the rest went off together and took a look at the far side of their nation at the far end of an era. What they recorded—and maybe especially what they failed to note—tells me something about their time, and mine. My pursuit of history has brought me back to myself, yearning in a crowd for my own idealized past, a picture-perfect country as big as all imagining.

SOUVENIRS

T*he next day*, in the Smithsonian archives, I turn pages of the Souvenir Album, the four-volume gift book of photographs that Harriman gave all his guests to commemorate their time together. Lots of pictures of mountains and glaciers. Of the guests wearing life jackets for a safety drill. Flowers. Harriman's bear. Seal hunters at Yakutat, whalers at Port Clarence. A herd of sea lions stampeding in a blur, down a rocky shore to water.

I look for a long time at a photo from Plover Bay, on the Russian coast. In the foreground two bare-headed, windblown Native people in fur parkas—a man and a girl—face left, almost in profile. Twenty feet or so behind them, two of the Harriman women, in shoe-length coats with pinched waists and padded shoulders, with white hats balanced on their coiffed heads like plates, face right. There is stone-strewn tundra underfoot, water behind, then an abrupt mountainside streaked with snow gullies. The photo was taken not by Curtis, who did all the developing and printing for the album, or the other professional photographer on the trip, but by Dall.

I try to imagine what Dall was hoping to capture here, in this juxtaposition of cultures, facing left and right, the distance between them, the one at home, the other looking so exotically and impractically out of place. It could be that the Harriman women were his subject, the others simply a foreground to their adventure, as the mountains were background. Or, the Native people—close enough to see the details of their clothing, the man's aquiline nose and thin mustache—might have been more central to his interest. Dall, with all his expertise in paleontology, knew a fossil when he found one.

Then I remember, earlier, at the Smithsonian's Native Peoples exhibit, the life-size diorama of the fur-clad Eskimo family hunting a seal, the Japanese tourists taking turns posing in front of it, as though they were part of the scene, as though they were members of the family. The fake Eskimos were caricatures, frozen on a Nanook-of-the-North stage, and the tourists were grinning at their own cleverness.

It is a shame, maybe, that Curtis went on from the expedition to do the kind of work he became famous for—photographing Native Americans not as they really were but only in a romanticized vision of how they had been, posed in ceremonial clothes with their drums and arrows. Maybe, because we lost the true documentation he might have given us. But maybe not, because he gave us that other thing, the story of what his age wanted to think about Native Americans, the myth-making of noble savages.

That is not so different from what, through his long career, Burroughs gave us in language—his idyllic scenes and stories, all those lovely feel-good flowers.

Here is another photo, this one by Merriam. It is an interior house pole from Cape Fox Village, holding up a roof: a carved bear, grasping, in its front paws, a small person with a hideously wide-eyed face. The person, upside down, is held by its shoulders, and its body, below the naked chest, disappears into the bear's mouth. The person's long hair, gathered into a ponytail at the top of its head, hangs straight down over the bear's belly; this hair, not part of the carved wood, is real hair, hanging. The room around the pole looks trashy, trashed, light coming through a gap in the wall, containers tipped on their sides, scrap on the floor. I see as I hadn't before that this place was deteriorating, coming apart, that it had already, before the H.A.E. visit, been scavenged. Someone—Curtis? Harriman?—has titled this photo "A Household God." The title seems trivializing, seems to say *heathen superstition*, seems to ignore the real power and the beauty of the sculpture, its true mystery.

I come to Curtis's photo, titled "The Two Johnnies," of Burroughs and Muir, posed before rock and snow on Saint Matthew's Island. I can see in this sharp photo, as I never have in its reproductions, the perfect detail of the individual flowers in the bouquet Muir holds in one hand. Burroughs, facing Muir from a few feet away, is also holding something against his chest—not another bouquet, it seems, but something more solid, what looks like a clump of tundra. The breeze has parted Burroughs's coat, and his pointy-toe boots look firmly planted. I see clearly that, under his dark hat, hair curls about the bottom of his ear, and kindly lines crinkle at the corner of one eye. Muir is facing the camera, but Burroughs is facing Muir and, I think, saying something that softens his face, though his mouth is hidden behind beard.

Burroughs will write, at the end of his account, about coal left in the bunkers and the cow that had been to Siberia and back, giving milk all the way. He will forget his discomforts and embarrassment and put into print, "No voyagers were ever more fortunate than we. No storms, no winds, no delays nor accidents to speak of, no illness. We had gone far and fared well."

I page through the Souvenir Album, all four fat volumes. The landscapes are familiar, not much changed in a hundred years, though modern photography can trick them into more dramatic shape and color. The wildlife shots aren't much—the dead bear, the stampeding sea lions, those fuzzy baby geese on the boat's deck. What I linger over again and again is the meager record of Alaska's human life—that massive church at Metlakatla, a group of children at Port Clarence, the house pole, my town of Homer as it was. I strain after details, the pair of mukluks and liners lying to dry on a beach in the foreground of one photo, the way sealskins were stretched around wooden hoops with crossbars, the frightened face of a child sucking her hand. There are no photos here of fox farms, canneries, mines—those things that were not, I suppose, scenic, the "souvenir" kind.

The photographers had an assignment. The scientists had theirs. Burroughs had his writing task. I don't suppose any of them knew what a fleeting moment they had found in Alaska, what a time they had entered into, what would soon be lost. Or gained. They would be surprised, I think, to see bears everywhere along the coast today, and sea otters, and Native men and women hauling in full nets of shining salmon. *I'm* surprised, knowing how close we came to sending more large species the way of the sea cow, to devastating the land I love. The greater casualties were people, cultures, those unique, ancient, finely tuned ways of living that were barely even glimpsed until it was too late.

At five o'clock, I close the last heavy photo binder. The woman at the desk is watching me over the tops of her glasses, waiting to sign me out.

I see now that what I have tracked has been a story of our nation, of its westering impulse and obsession, its search for a home place, a chase after the myth that wherever we go we will make what we find our own. Burroughs could have his brief fantasy about settling on a Kodiak fur farm, I that I might forever find a farther and more perfect, unspoiled west. We *have* traveled far and, yes, most of us have fared well. Although the blank spaces on our maps have filled in, one by one, and our arrogance has left chaos and ruin along our paths, we might yet know that the one world we have, in all its scope and variety, is the only gift we get.

Soon I will be aboard a jet, and then coasting in low over Alaska's Chugach Mountains, all sharp white contours and windblown wildness. That is the vision I carry with me: Alaska in winter, with the promise of green, and an idea that we might all someday manage to live in places we love, where we honor the many ways of knowing. The jet will drop toward Cook Inlet, the ice beside Anchorage fitting into exquisite hexagonal patterns, and my chest will choke again with what matters in the end, century to century—the hope and beauty of it all.